The Mathematics
Survival Kit

Jack Weiner

University of Guelph

THOMSON

NELSON

Australia   Canada   Mexico   Singapore   Spain   United Kingdom   United States

THOMSON

NELSON

The Mathematics Survival Kit
by Jack Weiner

Editorial Director and Publisher:
Evelyn Veitch

Acquisitions Editor:
Anthony Rezek

Marketing Manager:
Don Thompson

Managing Production Editor:
Susan Calvert

Proofreader:
Edie Franks

Production Manager:
Renate McCloy

Creative Director:
Angela Cluer

Cover Design:
Peter Papayanakis

Cover Image:
Peter Papayanakis

Printer:
Webcom

National Library of Canada Cataloguing in Publication Data

Weiner, Jack, 1949–
    The mathematics survival kit / Jack Weiner.

Includes index.
ISBN 0-17-641618-8

    1. Algebra—Outlines, syllabi, etc. 2. Calculus—Outlines, syllabi etc. I. Title.

QA37.3.W44 2003   512'.15
C2003-904111-5

THE MATHEMATICS SURVIVAL KIT: TABLE OF CONTENTS

Introduction

Getting Started on Survival

Factoring: A Product of Practice

Powerful with Polynomial Expressions

The Straight Goods on Lines and Planes

A Few Lines on Linear Algebra

Quadratics to the Second Degree!

Angling in on Trigonometry and a Little Geometry Too!

A Straightforward Approach to Limits

Continuity (There's a Hole in the Function, Dear Liza, Dear Liza)

Introducing "The Mathematics Survival Kit"

It's 2 a.m. and you are stuck. You have forgotten how to "complete the square", which you need to complete a calculus question or an algebra question or a statistics question or ... Your text gives lots of examples but none of them review the completing the square technique.

Here comes **The Mathematics Survival Kit** to the rescue! Find the page you need in the **Survival Kit**. A quick 5 minute concise and friendly review gives you exactly what you need to continue with your homework.

The Mathematics Survival Kit contains **115** topics just like that. Using high school course outlines, standard first year university calculus and algebra curricula, and my thirty years of teaching experience, I prepared a list of topics, from early high school to university, that plague students. Like completing the square. Like factoring cubics. Like those annoying log properties. Like ...

Each topic is covered in one page. Almost all consist of
   ➤ an introductory sentence or two
   ➤ a first very straightforward example
   ➤ a second example illustrating a common variation
   ➤ **TWO FOR YOU**: two extra questions with answers to reinforce the techniques (and space in between for you to **do the work!**)

I have three subtitles for this book and I believe in all three to the depths of my mathematics teacher's soul:

"Got a Math Problem? Give Me Five Minutes of Your Time!"

"I Remember How to Do That!"

"That's Easier Than I Remembered!"

Why **"Give me Five Minutes!"**? Pre-calculus and math review texts re-teach and do massive amounts of examples and exercises. They can be overwhelming. With the **Survival Kit**, I am saying, "Give me five minutes and together we will zero in on and solve **your** problem!" Also, pre-calculus math review texts miss many of these topics. I don't think there is any book available that covers them all because the topics span so much of the math curriculum. There is certainly no text out there that presents them in this one page per topic format.

That is a key point. You don't have to read pages 1 to 58 to understand page 59. Each page is like a dictionary definition. You have a vocabulary that allows you to understand the explanation of any word you look up in a dictionary. You have a mathematics vocabulary that does the same for **Mathematics Survival Kit** topics!

Why **"I Remember How to Do That!"**? You are smarter and know more than you often give yourself credit for. If I have done my job well and you have studied the material in an earlier course, you should be able to follow step by step the solutions to examples on any given page.

Why **"That's Easier Than I Remembered!"**? If you are using this book, you are probably strongly motivated and really into the math (or physics or chemistry or economics or ...) that you are currently studying. You have the background and motivation. You **can** do math. And most of the time, you will find that the topic you are reviewing is not nearly as intimidating as it once may have seemed.

I had a lot of fun writing this book and truly believe in its usefulness. Students, from **early high school to senior university**, who have seen drafts have said they want copies ASAP. Teachers have said it will save them loads of tutoring time because the extra help their students often need is background review – **The Survival Kit**! – rather than the current material. Parents have all said they want it not only for their offspring, but – I am not making this up – because it will be fun to see how much math they remember. Maybe enough to help junior!

Three people proofed and proofed and blew the errors down.

Thank you, Liisa Lahtinen. Liisa is a Masters of Mathematics student at the University of Guelph and one of my best Teaching Assistants ever.

Thank you, Professor Herb Kunze. Herb is a colleague in my department and one of the most helpful, giving people I have ever been privileged to know. He is also (this could get me in trouble) the best teacher in the department. Well, Gerarda's really good too. So is Joe. And Hosh ... Okay, Herb is one of the best teachers in the department.

I extend my gratitude to Professor Hosh Pesotan, who gave me excellent suggestions as the book progressed and continually enthused, "**The Mathematics Survival Kit** is such a neat idea. Why hasn't anyone done it before?"

Thank you, Anthony Rezek, Nelson Acquisitions Editor. Anthony nurtured both me and **The Mathematics Survival Kit** through the writing process, patiently and promptly offering advice on all my editing questions.

Thank you, Matthew Fisher. Matthew is a senior undergraduate computer science major and one of my best Student Assistants ever. The index is 99% Matthew!

Finally, and most importantly, I am grateful to the literally tens of thousands of students I have had the honour to teach here at the University of Guelph and, in a previous geological age, at Parkside High School in Dundas, Ontario. I hope I have made a positive contribution to many of their lives. They certainly have enriched mine.

Of course, any imperfections that remain (and as the former editor of a mathematics journal, I know that achieving perfection is an "asymptotic" limit!) are mine and mine alone.

Jack Weiner
Associate Professor
Department of Mathematics and Statistics
University of Guelph
June, 2003

ABOUT THE AUTHOR

Professor Jack Weiner taught at the University of Guelph from 1974 to 1976. He spent the next five years at Parkside High School in Dundas, Ontario. In 1982, he was re-recruited by Guelph and has been happily teaching and writing there ever since. He has won both the University of Guelph Professorial Teaching Award and the prestigious Ontario Association of University Faculty Association Teaching Award. He has been listed as a "Popular Professor" in MacLean's annual Canada wide University survey five years out of six!

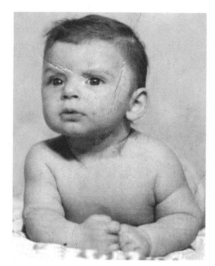

Jack showed a keen interest in mathematics from the very beginning. Here he is at age three months contemplating a math problem, which, thirty seconds later… he solved!

Jack in class today.

HOW TO GET AN "A" IN MATH!

1) After class, **DON'T** do your homework! Instead, read over your class notes. When you come to an example done in class...

2) **DON'T** read the example. Copy out the question, set your notes aside, and do the question yourself. Maybe you will get stuck. Even if you thought you understood the example completely when the teacher went over it in class, you may get stuck.

And this is **GOOD NEWS**! Now, you know what you don't know. So, consult your notes, look in the text, see your teacher/professor. Do whatever is necessary to figure out the steps in the example that troubled you.

Once you have sweated through the example, **DO IT AGAIN! And again**. Do it as often as you need so that it becomes, if not easy, then at least straightforward. Make sure you not only understand each line in the solution, but why each line is needed for the solution.

In part, you have memorized the solution. More importantly, you have made the subtleties of the problem unsubtle!

This is the great equalizer step. If your math or science aptitude is strong, then maybe you will have the example down pat after doing it twice. If not so strong, you may have to do it several times. But after you have done this for every class example...

3) **DO YOUR HOMEWORK!**

If you follow this method and if the teacher chose the examples well, then most of the homework questions will relate easily back to problems done in class and the rest should extend or synthesize the ideas behind those problems.

Guess what you'll find on 80% or more of your tests and exams? The same kinds of problems! And you will have your "A". Good luck, although if you use this method, luck will have nothing to do with your INEVITABLE success.

HOW TO GET EXTRA HELP

Have your questions ready. When you see your teacher for extra help, **don't say anything like**, "I don't have a clue what's going on." Rather, work through your class notes – definitions, examples, theorems and proofs – thoroughly, and be prepared to say, "I understand everything up to this point. How did we get from here to here?" In other words, **do your part!** Spend quality time with the material.

Get into study groups. Then one representative of your group can see your teacher to help with problems and report back to the others. Also, with group expertise, more often than not, you will solve most problems yourselves.

Struggle more than a little. Don't give up after one attempt. Make a sincere effort to sort out your problems. That way, when you say, "I am stuck RIGHT HERE!", you will be so ^{up} on the problem that your teacher's words of wisdom (I am not being sarcastic here!) won't seem like a foreign language.

Have you ever had that experience? A teacher is explaining something as if it is the most straightforward concept imaginable. You are nodding your head, saying, "Uh huh! Uh huh! Got it." And you're thinking, "I don't have a clue about what this teacher is talking about." It's rare, but sometimes this **is** the teacher's fault. Often, though, the blame lies with the student, who hasn't struggled enough so that the teacher's explanation can work. **Struggle more than a little!**

SOME SPECIFIC STRATEGIES

‣Review each day's notes as soon as possible – definitely before the next class.

‣You should, on average, do a **minimum** of one hour of homework per class.

‣**Participate in class.**

‣Form study groups with classmates.

‣Don't fall behind.

Factoring Difference of Squares

Does anyone out there have a problem with $a^2 - b^2 = (a-b)(a+b)$? I didn't think so!

Problems arise when it's not so obvious that difference of squares is what we are dealing with.

Example 1) Factor (a) $a^4 - b^8$ (b) $(x+y-z)^2 - (x-y-z)^2$

Solution (a) $a^4 - b^8 = (a^2 - b^4)(a^2 + b^4) \overset{\boxed{\text{Don't stop now!}}}{=} (a - b^2)(a + b^2)(a^2 + b^4)$

$\overset{\boxed{\text{for the obsessive}}}{=} (\sqrt{a} - b)(\sqrt{a} + b)(a + b^2)(a^2 + b^4)$

(b) $(x+y-z)^2 - (x-y-z)^2 \overset{\boxed{\substack{a = x+y-z \\ b = x-y-z}}}{=} \big(x+y-z-(x-y-z)\big)\big(x+y-z+x-y-z)\big)$

$= 2y(2x - 2z) = 4y(x - z)$

Difference of squares is often used **SDRAWKCAB**, er, **BACKWARDS**, to rationalize expressions This is especially useful in some limit questions.

Example 2) Rationalize the denominator in $\dfrac{1}{\sqrt{x} + \sqrt{y}}$.

Solution $\dfrac{1}{\sqrt{x} + \sqrt{y}} = \dfrac{1}{\sqrt{x} + \sqrt{y}}\left(\dfrac{\sqrt{x} - \sqrt{y}}{\sqrt{x} - \sqrt{y}}\right) = \dfrac{\sqrt{x} - \sqrt{y}}{x - y}$

Example 3) Evaluate: $\displaystyle\lim_{x \to 9} \dfrac{x - 9}{\sqrt{x} - 3}$

First Solution $\displaystyle\lim_{x \to 9} \dfrac{x - 9}{\sqrt{x} - 3} \overset{\boxed{\text{Factor the top...}}}{=} \lim_{x \to 9} \dfrac{(\sqrt{x} - 3)(\sqrt{x} + 3)}{\sqrt{x} - 3} = \lim_{x \to 9}(\sqrt{x} + 3) = 6$

Second Solution $\displaystyle\lim_{x \to 9} \dfrac{x - 9}{\sqrt{x} - 3} \overset{\boxed{\text{...or rationalize the bottom.}}}{=} \lim_{x \to 9} \dfrac{x - 9}{(\sqrt{x} - 3)}\left(\dfrac{\sqrt{x} + 3}{\sqrt{x} + 3}\right)$

$= \displaystyle\lim_{x \to 9} \dfrac{(x - 9)(\sqrt{x} + 3)}{x - 9} = \lim_{x \to 9}(\sqrt{x} + 3) = 6$

Two for you.

1) Factor: (a) $(x+y)^2 - (x-y)^2$ (b) $x^4 - z^{12}$

2)(a) Rationalize the numerator: $\dfrac{\sqrt{x+4} - \sqrt{3x-6}}{x-5}$ (b) Evaluate: $\displaystyle\lim_{x\to 5} \dfrac{\sqrt{x+4} - \sqrt{3x-6}}{x-5}$

Answers 1)(a) $4xy$ (b) $(x - z^3)(x + z^3)(x^2 + z^6)$

2)(a) $\dfrac{-2}{\sqrt{x+4} + \sqrt{3x-6}}$ (b) $-\dfrac{1}{3}$

Factoring Difference of Cubes

Lots of people have trouble with this one!

$a^3 - b^3 = (a-b)(a^2 + ab + b^2)$. Note that the coefficient of ab is +1. Note also that $a^2 + ab + b^2$ doesn't factor any further!

Example 1) Factor (a) $a^3 - b^6$ (b) $(x+h)^3 - x^3$

Solution (a) $a^3 - b^6 \overset{\boxed{a^3 - \left(b^2\right)^3}}{=} (a - b^2)(a^2 + ab^2 + b^4) \overset{\boxed{\text{For the obsessive: use difference of squares.}}}{=} (\sqrt{a} - b)(\sqrt{a} + b)(a^2 + ab^2 + b^4)$

(b) $(x+h)^3 - x^3 \overset{\boxed{\text{Here, } a = x+h \text{ and } b = x.}}{=} (x+h-x)\left((x+h)^2 + (x+h)x + x^2\right) = h\left((x+h)^2 + (x+h)x + x^2\right)$

(If you have taken calculus, this example should remind you of the derivative limit for $y = x^3$.)

As with difference of squares, there are two approaches to limits with difference of cubes.

Example 2) Evaluate $\displaystyle \lim_{x \to 64} \frac{x - 64}{x^{1/3} - 4}$.

First Solution $\displaystyle \lim_{x \to 64} \frac{x - 64}{x^{1/3} - 4} \overset{\boxed{\substack{\text{Factor the top:} \\ a = x^{1/3} \text{ and } b = 4}}}{=} \lim_{x \to 64} \frac{(x^{1/3} - 4)(x^{2/3} + 4x^{1/3} + 16)}{x^{1/3} - 4}$

$\displaystyle = \lim_{x \to 64} (x^{2/3} + 4x^{1/3} + 16) = 48$

Second solution $\displaystyle \lim_{x \to 64} \frac{x - 64}{x^{1/3} - 4} \overset{\boxed{\text{Now rationalize the bottom.}}}{=} \lim_{x \to 64} \frac{x - 64}{(x^{1/3} - 4)} \left(\frac{x^{2/3} + 4x^{1/3} + 16}{x^{2/3} + 4x^{1/3} + 16} \right)$

$\displaystyle = \lim_{x \to 64} \frac{(x - 64)(x^{2/3} + 4x^{1/3} + 16)}{x - 64} = \lim_{x \to 64} (x^{2/3} + 4x^{1/3} + 16) = 48$

Two for you.

1) Factor $x^{12} - y^{12}$.

2) Evaluate the limits: (a) $\lim\limits_{x \to 5} \dfrac{x^3 - 125}{x^2 - 25}$ (b) $\lim\limits_{x \to 8} \dfrac{x - 8}{x^{1/3} - 2}$

Answers 1) $(x^4 - y^4)(x^8 + x^4 y^4 + y^8) = (x - y)(x + y)(x^2 + y^2)(x^8 + 4x^4 y^4 + y^8)$

2)(a) $\dfrac{15}{2}$ (b) 12

Factoring $a^n - b^n$ and $a^n + b^n$

First: Factoring $a^n - b^n$

First, you might want to look in the **Survival Kit** for **difference of squares** and **difference of cubes**.

$$a^2 - b^2 = (a-b)(a+b)$$
$$a^3 - b^3 = (a-b)(a^2 + ab + b^2) \text{ Note that the coefficient of } ab \text{ is 1.}$$
$$a^4 - b^4 = (a-b)(a^3 + a^2 b + ab^2 + b^3). \text{ So, for positive integers } n,$$
$$a^n - b^n = (a-b)(a^{n-1} + a^{n-2} b + a^{n-3} b^2 + a^{n-4} b^3 + \dots + a^2 b^{n-3} + ab^{n-2} + b^{n-1})$$

In the second bracket for the factored form of $a^n - b^n$, the exponent on a starts at $n-1$ and decreases one by one down to 0. The exponent on b starts at 0 and goes up one by one to $n-1$.
THIS FORMULA WORKS FOR ANY NATURAL NUMBER n.

Second: Factoring $a^n + b^n$
(and we want n to be ODD!)

Face it: $a^2 + b^2$ **doesn't factor!** Well, all right, it does **IF YOU ALLOW COMPLEX NUMBERS**. $a^2 + b^2 = (a - bi)(a + bi)$, where $i = \sqrt{-1}$. For our purposes, restricted to real numbers, the sum of squares does not factor. Neither does $a^4 + b^4$ or $a^8 + b^8$. However, sum of cubes definitely does: $a^3 + b^3 = (a+b)(a^2 - ab + b^2)$? Compare this to the difference of cubes.
$a^3 - b^3 = (a-b)(a^2 + ab + b^2)$ Look at where + changes to −.
So, as long as n is $\boxed{\text{ODD!!}}$,
$$a^n + b^n = (a+b)(a^{n-1} - a^{n-2} b + a^{n-3} b^2 - a^{n-4} b^3 + \dots + a^2 b^{n-3} - ab^{n-2} + b^{n-1})$$

Example 1) Factor: (a) $a^5 - b^5$ (b) $a^5 + b^5$ (c) $a^7 + b^7$

Solution (a) $a^5 - b^5 = (a-b)(a^4 + a^3 b + a^2 b^2 + ab^3 + b^4)$
(b) $a^5 + b^5 = (a+b)(a^4 - a^3 b + a^2 b^2 - ab^3 + b^4)$
(c) $a^7 + b^7 = (a+b)(a^6 - a^5 b + a^4 b^2 - a^3 b^3 + a^2 b^4 - ab^5 + b^6)$

Two for you.

1) Factor $a^5 - b^{10}$.

2) Factor $a^{15} + b^{30}$.   (Hint: $a^{15} + b^{30} = (a^3)^5 + (b^6)^5$)

Answers 1) $(a - b^2)(a^4 + a^3 b^2 + a^2 b^4 + ab^6 + b^8)$

2) $(a^3 + b^6)(a^{12} - a^9 b^6 + a^6 b^{12} - a^3 b^{18} + b^{24}) = (a + b^2)(a^2 - ab^2 + b^4)(a^{12} - a^9 b^6 + a^6 b^{12} - a^3 b^{18} + b^{24})$

The Remainder and Factor Theorems for Polynomials

> **The Remainder Theorem** tells us that when we divide polynomial $P(x)$ by $x - a$ to obtain quotient $Q(x)$ and remainder R, then $R = P(a)$, that is, $P(x) = Q(x)(x - a) + P(a)$.
> So, if $P(a) = 0$, then $P(x) = Q(x)(x - a)$, that is, $x - a$ is a factor of $P(x)$. This is **The Factor Theorem!** Also, suppose $P(x) = a_n x^n + a_{n-1} x^{n-1} + \ldots + a_1 x + a_0$, where all the a_i's are integers, and $\dfrac{p}{q}$ is a root of $P(x)$. **Then q must divide a_n and p must divide a_0.**

Example 1) Use the Factor Theorem to find the **rational roots** of $P(x) = x^3 - x^2 - 4x + 4$.

Solution The coefficient of x^3 is 1 and $a_0 = 4$. If $a = \dfrac{p}{q}$ is a rational root, then

p must divide 4 and q must divide 1. Therefore, we only need to check $a = \pm 1, \pm 2, \pm 4$.
$P(1) = 0, \ P(-1) = 6, \ P(2) = 0, \ P(-2) = 0, \ P(4) = 36, P(-4) = -60$
Therefore the roots are 1, 2, and -2.

Two notes : 1) $P(x)$, a cubic, can have at most three roots. So, if we find three, we are definitely done!
         2) $P(x) = (x - 1)(x - 2)(x + 2)$

Example 2) Find the rational roots of $P(x) = 2x^4 - 5x^3 + 5x^2 - 5x + 3$.

Solution The coefficient of x^4 is 2 and $a_0 = 3$. If $a = \dfrac{p}{q}$ is a rational root, then

p must divide 3 and q must divide 2. Therefore, we need to check $a = \pm 1, \pm \dfrac{1}{2}, \pm 3$, and $\pm \dfrac{3}{2}$.

$P(1) = 0, \ P(-1) = 20, \ P(3) = 60, \ P(-3) = 360, \ P\left(\dfrac{1}{2}\right) = \dfrac{5}{4}, \ P\left(-\dfrac{1}{2}\right) = \dfrac{15}{2}, \ P\left(\dfrac{3}{2}\right) = 0, \ P\left(-\dfrac{3}{2}\right) = \dfrac{195}{4}$

Therefore the rational roots are 1 and $\dfrac{3}{2}$.

Two notes : 1) Since $P(x)$ is degree 4, the remaining roots are either irrational or complex.

         2) In fact, by dividing $x - 1$ into $P(x)$ and then $x - \dfrac{3}{2}$ into the resulting quotient,

         you would find $P(x) = (x - 1)\left(x - \dfrac{3}{2}\right)(2x^2 + 2) \ \overset{\boxed{\text{a little prettier...}}}{=} \ (x - 1)(2x - 3)(x^2 + 1)$.

Two for you.

Find the rational roots and then factor : 1) $P(x) = x^3 + 2x^2 - 5x - 6$ 2) $P(x) = 3x^4 - x^3 - 3x + 1$

Answers 1) $-1,\ 2,\ -3,\ P(x) = (x+1)(x-2)(x+3)$ 2) $1,\ \dfrac{1}{3},\ P(x) = (3x-1)(x-1)(x^2 + x + 1)$

Adding and Subtracting Polynomial Fractions

To simplify when we add or subtract fractions, we need to get the

lowest common denominator : $\dfrac{2}{3}+\dfrac{5}{9}-\dfrac{5}{12}=\dfrac{24}{36}+\dfrac{20}{36}-\dfrac{15}{36}=\dfrac{29}{36}$.

We use exactly the same method when adding and/or subtracting fractions with polynomials in the top and bottom.

Example 1) Simplify $\dfrac{3}{2a}-\dfrac{6}{5a}+\dfrac{3}{10a}$.

Solution $\dfrac{3}{2a}-\dfrac{6}{5a}+\dfrac{3}{10a}$

Get the **lowest** common denominator!

$=\dfrac{15}{10a}-\dfrac{12}{10a}+\dfrac{3}{10a}$

Simplify the numerator.

$=\dfrac{6}{10a}$

Reduce more if you can!

$=\dfrac{3}{5a}$

Example 2) Simplify $\dfrac{2x+1}{x-1}-\dfrac{x+1}{x+2}-\dfrac{5x+4}{x^2+x-2}$.

Solution $\dfrac{2x+1}{x-1}-\dfrac{x+1}{x+2}-\dfrac{5x+4}{x^2+x-2}$

Factor the denominators!

$=\dfrac{2x+1}{x-1}-\dfrac{x+1}{x+2}-\dfrac{5x+4}{(x-1)(x+2)}$

Get the **lowest** common denominator.

$=\dfrac{(2x+1)(x+2)}{(x-1)(x+2)}-\dfrac{(x+1)(x-1)}{(x-1)(x+2)}-\dfrac{5x+4}{(x-1)(x+2)}$

Expand the numerator.

$=\dfrac{2x^2+5x+2-(x^2-1)-(5x+4)}{(x-3)(x+3)(x-2)}$

Now simplify the numerator.

$=\dfrac{x^2-1}{(x-1)(x+2)}$

Check for and divide out any further common factors.

$=\dfrac{\cancel{(x-1)}(x+1)}{\cancel{(x-1)}(x+2)}=\dfrac{x+1}{x+2}$

Two for you.

Simplify each of the following rational expressions:

1) $\dfrac{2x}{x-5} - \dfrac{x}{x-3} + \dfrac{1}{x^2-8x+15}$

2) $\dfrac{x-1}{x^2-16} + \dfrac{x}{x^2-5x+4} - \dfrac{1}{x^2+3x-4}$

Answers 1) $\dfrac{x^2-x+1}{(x-5)(x-3)}$

2) $\dfrac{2x^2+x+5}{(x-4)(x+4)(x-1)}$

Multiplying and Dividing Polynomial Fractions

What we do with fractions having polynomials in the top and bottom is **exactly** what we do with fractions having numbers in the top and bottom.

$$\frac{4\cdot 7^3}{2\cdot 5^5}\times\frac{2^4\cdot 5^4}{4^2\cdot 7\cdot 11}\;\boxed{\text{Get common factors and bases.}}\;=\;\frac{2^2\cdot 7^3\cdot 2^4\cdot 5^4}{2\cdot 5^5\cdot 2^4\cdot 7\cdot 11}\;\boxed{\text{Divide out the common factors.}}\;=\;\frac{2\cdot 7^2}{5\cdot 11}\;\boxed{\text{In the numerical case, work out the final value.}}\;=\;\frac{98}{55}\;\boxed{\text{or}}\;=1\frac{43}{55}$$

Let's do the same question, setting it up so that it starts as **DIVISION!**

$$\frac{4\cdot 7^3}{2\cdot 5^5}\div\frac{4^2\cdot 7\cdot 11}{2^4\cdot 5^4}\;\boxed{\text{Invert and multiply!}}\;=\;\frac{4\cdot 7^3}{2\cdot 5^5}\times\frac{2^4\cdot 5^4}{4^2\cdot 7\cdot 11}\;\boxed{\text{Now repeat the steps above.}}\;=\;\ldots=\frac{98}{55}$$

Example 1) Simplify the rational expression $\dfrac{(x^2-9)}{(x-3)^3}\times\dfrac{(x^2-3x)^2}{x^3+27}$.

Solution $\dfrac{(x^2-9)}{(x-3)^3}\times\dfrac{(x^2-3x)^2}{x^3+27}$

$\boxed{\text{Factor first! Then divide out the common factors.}}$
$$=\frac{(\cancel{x-3})(\cancel{x+3})(x^2)(\cancel{x-3})^2}{(\cancel{x-3})^3(\cancel{x+3})(x^2-3x+9)}$$

$$=\frac{x^2}{x^2-3x+9}$$

Example 2) Simplify the rational expression $\dfrac{m^2+3mn+2n^2}{m^2+2mn+n^2}\div\dfrac{m^2+2mn}{m^2+mn}$.

Solution $\dfrac{m^2+3mn+2n^2}{m^2+2mn+n^2}\div\dfrac{m^2+2mn}{m^2+mn}$

$\boxed{\text{Invert and multiply!}}$
$$=\frac{m^2+3mn+2n^2}{m^2+2mn+n^2}\times\frac{m^2+mn}{m^2+2mn}$$

$\boxed{\text{Factor and divide out common factors!}}$
$$=\frac{(\cancel{m+n})(\cancel{m+2n})}{(\cancel{m+n})^2}\times\frac{\cancel{m}(\cancel{m+n})}{\cancel{m}(\cancel{m+2n})}$$

$$=1$$

Two for you.

Simplify each of the following rational expressions:

1) $\dfrac{(x^2+5x+6)^2}{(x^2+6x+9)(x^2-9)} \times \dfrac{x^3-27}{x^2+4x+4}$

2) $\dfrac{(a^3+5a^2)^3}{a^2+10a+25} \div \dfrac{a^8(a^3+125)}{a^2-5a+25}$

Answers   1) $\dfrac{x^2+3x+9}{x+3}$ 2) $\dfrac{1}{a^2}$

14

Polynomial Division

Keep this example in mind: $3\overline{\smash{)}28}$ with $\dfrac{9}{}$ above, $\underline{-27}$, 1

| 3 divides into 28 approximately 9 times. |
| Multiply 9 by 3 and subtract to get the remainder. |

$$\frac{DIVIDEND}{divisor} = QUOTIENT + \frac{REMAINDER}{divisor} \quad \text{and} \quad DIVIDEND = QUOTIENT \cdot divisor + REMAINDER$$

$$\frac{28}{3} \quad = \quad 9 \quad + \quad \frac{1}{3} \quad \text{and} \quad 28 \quad = \quad 9 \cdot 3 \quad + \quad 1$$

Example 1) Divide the polyonomial $x^3 - 2x^2 + 5x - 7$ by $x + 2$.

Solution $x + 2$ divides into $x^3 - 2x^2 + 5x - 7$ approximately x^2 times. (Just like "3 into 28" above.)

Multiply $x + 2$ by x^2 and subtract to get the remainder. AND SO ON...

$$
\begin{array}{r}
x^2 - 4x + 13 \\
x+2{\overline{\smash{)}\,x^3 - 2x^2 + 5x - 7}} \\
\underline{-(x^3 + 2x^2)} \\
-4x^2 + 5x - 7 \\
\underline{-(-4x^2 - 8x)} \\
13x - 7 \\
\underline{-(13x + 26)} \\
-33
\end{array}
$$

Therefore, $\dfrac{x^3 - 2x^2 + 5x - 7}{x + 2} = x^2 - 4x + 13 - \dfrac{33}{x+2}$ $\left(\dfrac{D}{d} = Q + \dfrac{R}{d}\right)$

Alternatively, $x^3 - 2x^2 + 5x - 7 = (x^2 - 4x + 13)(x + 2) - 33$ $\left(D = Q \cdot d + R\right)$

Example 2) Divide the polyonomial $x^3 + 5x - 1$ by $x - 1$.

Solution For convenience, let's write $x^3 + 5x - 1 = x^3 + 0x^2 + 5x - 1$.

$$
\begin{array}{r}
x^2 + x + 6 \\
x-1{\overline{\smash{)}\,x^3 + 0x^2 + 5x - 1}} \\
\underline{-(x^3 - x^2)} \\
x^2 + 5x - 1 \\
\underline{-(x^2 - x)} \\
6x - 1 \\
\underline{-(6x - 6)} \\
5
\end{array}
$$

Therefore, $\dfrac{x^3 + 5x - 1}{x - 1} = x^2 + x + 6 + \dfrac{5}{x-1}$ $\left(\dfrac{D}{d} = Q + \dfrac{R}{d}\right)$

Alternatively, $x^3 + 5x - 1 = (x^2 + x + 6)(x - 1) + 5$ $\left(D = Q \cdot d + R\right)$

Two For You.

Divide: 1) $x^3 + 2x^2 + 3x + 4$ by $x - 1$ 2) $x^4 - x^3 - 5x + 4$ by $x - 2$

Answers 1) $x^3 + 2x^2 + 3x + 4 = (x^2 + 3x + 6)(x - 1) + 10$

2) $x^4 - x^3 - 5x + 4 = (x^3 + x^2 + 2x - 1)(x - 2) + 2$

Finding the Equation of a Line

Given the slope and a point, or two points, there are **lots** of ways to find the equation of a line. The easiest method to cover all scenarios uses the equation
$y - y_1 = m(x - x_1)$, where m is the slope and (x_1, y_1) is a point on the line.

Example 1) Find the equation of the line with slope -3 through the point $(2, -5)$.

Solution $y - (-5) \overset{\boxed{x_1 = 2,\, y_1 = -5,\, m = -3}}{=} -3(x - 2)$

$y + 5 = -3x + 6$

$y = -3x + 1$

Example 2) Find the equation of the line passing through the points $(2, 7)$ and $(4, 12)$.

Solution The slope $m = \dfrac{y_2 - y_1}{x_2 - x_1} = \dfrac{12 - 7}{4 - 2} = \dfrac{5}{2}$. $\boxed{\text{Note: } \dfrac{y_1 - y_2}{x_1 - x_2} = \dfrac{7 - 12}{2 - 4} = \dfrac{5}{2} \text{ as well.}}$

Using $(2, 7)$ as (x_1, y_1):   OR   Using $(4, 12)$ as (x_1, y_1):

$y - 7 = \dfrac{5}{2}(x - 2)$ $y - 12 = \dfrac{5}{2}(x - 4)$

$y - 7 = \dfrac{5}{2}x - 5$ $y - 12 = \dfrac{5}{2}x - 10$

$y = \dfrac{5}{2}x + 2$ $y = \dfrac{5}{2}x + 2$

Two for you.

1) Find the equation of the line through $(-1, -5)$ and $(3, 7)$.

2) Find the equation of the line with x intercept 4 and y intercept 7 (Hint: use the points $(4, 0)$ and $(0, 7)$.)

Answers 1) $y = 3x - 2$ 2) $y = -\dfrac{7}{4}x + 7$

Slope *m* and *y* intercept *b*

In the equation of the line $y = mx + b$, m is the slope and b is the y intercept.

Example 1) Find the slope and the y intercept for each of the following lines.

a) $y = 3x + 5$

b) $y = 7 - 2x$

c) $y = 4x$

d) $y = -7$

e) $2x + 3y = 4$

f) $4y - 5x - 2 = 0$

g) $x = 4$

h) $y = 0$

i) $x = 0$

Solution

a) $m = 3, b = 5$

b) $m = -2, b = 7$

c) $m = 4, b = 0$

d) $m = 0, b = -7$

e) Rewrite the equation in the form $y = mx + b$: $y = -\dfrac{2}{3}x + \dfrac{4}{3}$ and so $m = -\dfrac{2}{3}$ and $b = \dfrac{4}{3}$

f) $y = \dfrac{5}{4}x + \dfrac{1}{2}$ and so $m = \dfrac{5}{4}$ and $b = \dfrac{1}{2}$

g) The slope is undefined (or infinite). This is the vertical line where x **always equals 4** while y can be any real number. It is parallel to the y axis and **there is no y intercept**.

h) $m = b = 0$. ($y = 0$ is the equation of the x axis.)

i) The slope is undefined (or infinite). $x = 0$ is the equation of the y axis, so there are **LOTS** of y intercepts!

Two for you.

Find the slope and the y intercept for the lines

1) $\pi y - 2x = 1$ 2) $3x - 4 = 0$

Answers 1) $m = \dfrac{2}{\pi}$, $b = \dfrac{1}{\pi}$ 2) undefined slope, no y intercept

Distance between Two Points and
Distance from a Point to a Line or Plane

Example 1) Find the distance from $(-1, 5)$ to $(6, 3)$.

Solution Distance $= \sqrt{(x_2 - x_1)^2 + (y_2 - y_1)^2} = \sqrt{(6 - (-1))^2 + (3 - 5)^2} = \sqrt{7^2 + (-2)^2} = \sqrt{53}$

Example 2) Find the (perpendicular) distance from the point $(3, 4)$ to the line $4x - 5y = 7$.

Solution The distance from (x_0, y_0) to $Ax + By + C = 0$ is given by the formula $\dfrac{|Ax_0 + By_0 + C|}{\sqrt{A^2 + B^2}}$

and so, from the point $(3, 4)$ to the line $4x - 5y = 7$,

distance $\overset{\boxed{A = 4,\, B = -5,\, \text{and } C = -7}}{=} \dfrac{|(4)(3) + (-5)(4) - 7|}{\sqrt{4^2 + 5^2}}$

$= \dfrac{|-15|}{\sqrt{41}} = \dfrac{15}{\sqrt{41}}.$

Example 3) Find the distance from $(3, 4, -1)$ to $(0, 5, -2)$.

Solution Distance $= \sqrt{(x_2 - x_1)^2 + (y_2 - y_1)^2 + (z_1 - z_2)^2} = \sqrt{(0 - 3)^2 + (5 - 4)^2 + (-2 - (-1))^2}$

$= \sqrt{(-3)^2 + (1)^2 + (-1)^2} = \sqrt{11}$

Example 4) Find the perpendicular distance from the point $(1, 2, 3)$ to the plane $2x + 4y - 5z = 7$.

Solution The distance from (x_0, y_0, z_0) to $Ax + By + Cz + D = 0$ is $\dfrac{|Ax_0 + By_0 + Cz + D|}{\sqrt{A^2 + B^2 + C^2}}$ and so

from the point $(1, 2, 3)$ to the plane $2x + 4y - 5z = 7$,

distance $\overset{\boxed{A = 2,\, B = 4,\, C = -5,\, \text{and } D = -7}}{=} \dfrac{|(2)(1) + (4)(2) + (-5)(3) - 7|}{\sqrt{2^2 + 4^2 + (-5)^2}}$

$= \dfrac{|-12|}{\sqrt{45}} = \dfrac{12}{\sqrt{45}} = \dfrac{12}{3\sqrt{5}} = \dfrac{4}{\sqrt{5}} \overset{\boxed{\text{for fans of } \textbf{BOB} \\ (\text{Back Of the Book!})}}{=} \dfrac{4\sqrt{5}}{5}$

21

Two for you.

1) Find the distance between $(-3, 0, 1)$ and $(5, 2, 0)$.

2) Find the distance from the point $(1, 2, 3)$ to the plane $3x - 7z = 0$.

(Hint: $A = 3$, $B = 0$, $C = -7$, and $D = 0$.)

Answers 1) $\sqrt{69}$ 2) $\dfrac{18}{\sqrt{58}} \overset{\boxed{\text{BOB!}}}{=} \dfrac{9\sqrt{58}}{29}$ $\boxed{\text{BOB} \equiv \text{BACK OF BOOK}}$

Visually Identifying Slopes of Lines

This page is all about visually recognizing the slope of a line. Assume the scales on the x and y axes are equal in all cases. Slope tells you how much y increases for a unit change in x. So a slope of 3 means if x **increases** by 1, then y **increases** by 3. If x goes up by 5, then y goes up by $3 \times 5 = 15$, that is, y goes up 3 times as much as x. A slope of -3 means that if x **increases** by 1, then y **decreases** by 3.

Given points (x_1, y_1) and (x_2, y_2), the slope of the segment joining these points is

$$m = \frac{y_2 - y_1}{x_2 - x_1} = \frac{y_1 - y_2}{x_1 - x_2} = \frac{\Delta y}{\Delta x}$$

Warning! Don't confuse the symbol "m" for slope with the lines labelled "m" below.

Positive Slope
y goes up
as x goes up.

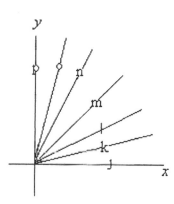

Slope p: undefined
Slope o $= 4$
Slope n $= 2$
Slope m $= 1$
Slope l $= 0.5$
Slope k $= 0.25$
Slope j $= 0$

Negative slope
y goes down
as x goes up.

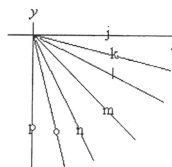

Slope j $= 0$
Slope k $= -0.25$
Slope l $= -0.5$
Slope m $= -1$
Slope n $= -2$
Slope o $= -4$
Slope p: undefined

Note:

Horizontal lines (line **j** in both pictures) have **slope 0**.

Vertical lines (line **p** in both pictures) have **undefined (or infinite) slope**.

One for you.

1) Match the slopes 3, -4, 0.5, -0.5 with the segments

(a)   (b)   (c)   (d)

Answer (a) -0.5 (b) 0.5   (c) -4 (d) 3

Parallel and Perpendicular Lines

If line l_1 with slope m_1 is parallel
to line l_2 with slope m_2, then $m_1 = m_2$.

If line l_1 with slope $m_1 \neq 0$ is perpendicular
to line l_2 with slope m_2, then $m_2 = -\dfrac{1}{m_1}$,
that is, the slope of l_2 is the **negative reciprocal** of the slope of l_1.

In the case of perpendicular lines, if $m_1 = 0$, then l_1 is
horizontal, that is, l_1 is parallel to the x axis. In this case,
l_2 is vertical, that is, parallel to the y axis, and its slope
is undefined, that is, l_2 has "infinite slope".

Example 1) Find the equation of the line which is (a) parallel   (b) perpendicular
to the line $y = -3x + 5$ which passes through the point $(-4, 3)$.

Solution (a) The slope of a parallel line is -3. Using $y - y_1 = m(x - x_1)$, we have
$y - 3 = -3(x - (-4))$ and so $y = -3x - 9$.

(b) The slope of a perpendicular line is $-\dfrac{1}{(-3)} = \dfrac{1}{3}$. Therefore,

$y - 3 = \dfrac{1}{3}(x + 4)$ and so $y = \dfrac{1}{3}x + \dfrac{13}{3}$.

Two for you.

1) Find the equation of the line through the point $(1, 1)$ that is

(a) parallel to the line through the points $(1, 1)$ and $(3, 11)$

(b) perpendicular to the line through the points $(1, 1)$ and $(3, 11)$.

2) Find the equation of the line with x intercept 1 that is

(a) parallel to $y = 4$ (b) perpendicular to $y = 4$.

Answers 1)(a) $y = 5x - 4$ (b) $y = -\dfrac{1}{5}x + \dfrac{6}{5}$

2)(a) $y = 0$ (Note: here, every real number is an x intercept!)   (b) $x = 1$

Finding Tangent and Normal Lines to a Curve

Let $y = f(x)$. The slope of the tangent line to this function at the point $(a, f(a))$ is given by $f'(a)$ and the slope of the normal line by $-\dfrac{1}{f'(a)}$.

Example 1) Let $f(x) = x^3 - 8x + 9$. Find the equation of the tangent and normal lines at the point where $x = 2$.

Solution $f'(x) = 3x^2 - 8$.

Now $f(2) = 1$ and $f'(2) = 4$. Using $y - y_1 = m(x - x_1)$, the tangent line is $y - 1 = 4(x - 2)$ and so $y = 4x - 7$. For the normal line, we still use the point $(2,1)$ but the slope is $-\dfrac{1}{4}$.

Therefore, the normal equation is

$y - 1 = -\dfrac{1}{4}(x - 2)$ and so

$y = -\dfrac{1}{4}x + \dfrac{1}{2} + 1 = -\dfrac{1}{4}x + \dfrac{3}{2}$.

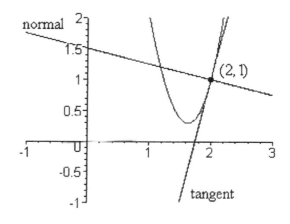

Example 2) Let $xy + e^{y-1} = 2$. Find the equation of the tangent and normal lines at the point where $y = 1$.

Solution Substituting $y = 1$ gives (remember $e^0 = 1$) $x + 1 = 2$ and so $x = 1$.

Differentiating implicitly: $x\dfrac{dy}{dx} + y + e^{y-1}\dfrac{dy}{dx} = 0$ $\therefore$ $\dfrac{dy}{dx}(x + e^{y-1}) = -y$ and so $\dfrac{dy}{dx} = \dfrac{-y}{x + e^{y-1}}$

At $(1,1)$, $\dfrac{dy}{dx} = -\dfrac{1}{2}$. The tangent line is $y - 1 = -\dfrac{1}{2}(x - 1)$ and so $y = -\dfrac{1}{2}x + \dfrac{3}{2}$.

The normal line is $y - 1 = 2(x - 1)$ and so $y = 2x - 1$.

Two for you.

1) Find the tangent and normal lines to $y = x^2 - x$ at $x = -1$.

2) Find the tangent and normal lines to $x + y + \ln y = 4$ at $y = 1$

Answers 1) tangent: $y = -3x - 1$ normal: $y = \dfrac{1}{3}x + \dfrac{7}{3}$

2) tangent: $y = -\dfrac{1}{2}x + \dfrac{5}{2}$ normal: $y = 2x - 5$

Solving Two Linear Equations

Let's use row reduction to solve these pairs of equations. There are other methods, and in fact some of these are easier than row reduction when you have only two equations, **but this is the easiest method to generalize to three, four, and more linear equations.**

Example Solve each of the following pairs of linear equations:

1) $x + 3y = 5$ and $4x - 5y = 3$ 2) $3x - 5y = 6$ and $5x - 8y = 0$

3) $2x - y = 7$ and $4x - 2y = 14$ 4) $2x - y = 7$ and $4x - 2y = 6$

Solution 1)

$$\begin{array}{|ll|} x + 3y = 5 & \text{E1} \\ 4x - 5y = 3 & \text{E2} \end{array}$$
$\overset{\boxed{\substack{\text{Leave E1 alone!} \\ -4 \times \text{E1} + \text{E2} = \text{E3}}}}{\Longleftrightarrow}$
$$\begin{array}{|ll|} x + 3y = 5 & \text{E1} \\ -17y = -17 & \text{E3} \end{array}$$

From $-17y = -17$, we find $y = 1$.

Substitute $y = 1$ into E1: $x + 3(1) = 5$ and so $x = 2$. The solution is $(x, y) = (2, 1)$.

2)
$$\begin{array}{|ll|} 3x - 5y = 6 & \text{E1} \\ 5x - 8y = 0 & \text{E2} \end{array}$$
$\overset{\boxed{\substack{\frac{1}{3} \times \text{E1} = \text{E3} \\ \text{Leave E2 alone!}}}}{\Longleftrightarrow}$
$$\begin{array}{|ll|} x - \dfrac{5}{3}y = 2 & \text{E3} \\ 5x - 8y = 0 & \text{E2} \end{array}$$

$\overset{\boxed{\substack{\text{Leave E3 alone!} \\ -5 \times \text{E3} + \text{E2} = \text{E4}}}}{\Longleftrightarrow}$
$$\begin{array}{|ll|} x - \dfrac{5}{3}y = 2 & \text{E3} \\ \dfrac{1}{3}y = -10 & \text{E4} \end{array}$$

From $\dfrac{1}{3}y = -10$, we have $y = -30$.

Substitute $y = -30$ into E1: $3x - 5(-30) = 6$ and so $3x = -144$ and $x = -48$.

3) Solving the same way as in 1) and 2) yields the equation $0 = 0$. These lines are **COINCIDENT** that is, both equations represent the same line. The solutions are $x \in \mathbb{R}$ and $y = 2x - 7$.

4) This time, we obtain the equation $0 = 8$. **FALSE!** There is no solution. These are non-intersecting parallel lines.

Two for you.

Solve the linear systems: 1) $5x - 7y = 10,\ 15x + 4y = 5$ 2) $4x + 3y = -1,\ 7x + 2y = 8$

Answers 1) $x = \dfrac{3}{5},\ y = -1$ 2) $x = 2,\ y = -3$

Solving Three Linear Equations

Use row reduction here just as with two equations in two unknowns. Long, tedious, easy to make mechanical errors, but the method is pretty straightforward.

Example Solve the following systems of linear equations:

1)
$$\begin{array}{l} x + 2y - 6z = 5 \\ 4x + 5y - 21z = 5 \\ -3x + 3y + 17z = -2 \end{array}$$

2)
$$\begin{array}{l} 2x + y - 2z = 10 \\ 3x + 2y + 2z = 1 \\ 5x + 4y + 3z = 4 \end{array}$$

Solution

1)

$$\begin{array}{ll} x + 2y - 6z = 5 & E1 \\ 4x + 5y - 21z = 5 & E2 \\ -3x + 3y + 17z = -2 & E3 \end{array}$$

Leave E1 alone!
$-4 \times E1 + E2 = E4$
$3 \times E1 + E3 = E5$
$\Leftrightarrow$

$$\begin{array}{ll} x + 2y - 6z = 5 & E1 \\ -3y + 3z = -15 & E4 \\ 9y - z = 13 & E5 \end{array}$$

Leave E1 alone!
$-\frac{1}{3} \times E4 = E6$
Leave E5 alone!
$\Leftrightarrow$

$$\begin{array}{ll} x + 2y - 6z = 5 & E1 \\ y - z = 5 & E6 \\ 9y - z = 13 & E5 \end{array}$$

Leave E1 alone!
Leave E6 alone!
$-9 \times E6 + E5 = E7$
$\Leftrightarrow$

$$\begin{array}{ll} x + 2y - 6z = 5 & E1 \\ y - z = 5 & E6 \\ 8z = -32 & E7 \end{array}$$

Therefore, $z = -4$
$y = 5 + z = 5 - 4 = 1$ and
$x = 5 - 2y + 6z = 5 - 2 - 24 = -21.$

2)

$$\begin{array}{ll} 2x + y - 2z = 10 & E1 \\ 3x + 2y + 2z = 1 & E2 \\ 5x + 4y + 3z = 4 & E3 \end{array}$$

$\frac{1}{2} \times E1 = E4$
Leave E2 alone!
Leave E3 alone!
$\Leftrightarrow$

$$\begin{array}{ll} x + y/2 - z = 5 & E4 \\ 3x + 2y + 2z = 1 & E2 \\ 5x + 4y + 3z = 4 & E3 \end{array}$$

Leave E4 alone!
$-3 \times E4 + E2 = E5$
$-5 \times E4 + E3 = E6$
$\Leftrightarrow$

$$\begin{array}{ll} x + y/2 - z = 5 & E4 \\ y/2 + 5z = -14 & E5 \\ 3y/2 + 8z = -21 & E6 \end{array}$$

Leave E4 alone!
$2 \times E5 = E7$
Leave E6 alone!
$\Leftrightarrow$

$$\begin{array}{ll} x + y/2 - z = 5 & E4 \\ y + 10z = -28 & E7 \\ 3y/2 + 8z = -21 & E6 \end{array}$$

Leave E4 alone!
Leave E7 alone!
$-\frac{3}{2} E7 + E6 = E8$
$\Leftrightarrow$

$$\begin{array}{ll} x + y/2 - z = 5 & E4 \\ y + 10z = -28 & E7 \\ -7z = 21 & E8 \end{array}$$

Therefore, $z = -3$
$y = -28 - 10z = -28 + 30 = 2$ and
$x = 5 - y/2 + z = 5 - 1 - 3 = 1.$

31

Two for you.

Solve the linear systems:

1) $x - 2y + z = 7,\ \ 2x - y + 4z = 17,\ \ 3x - 2y + 2z = 14$

2) $2x + y - 3z = 5,\ \ 3x - 2y + 2z = 5,\ \ 5x - 3y - z = 16$

Answers 1) $x = 2,\ y = -1,\ z = 3$ 2) $x = 1,\ y = -3,\ z = -2$

Consistent vs Inconsistent vs Dependent vs Unique Solutions of Three Linear Equations in 3 Unknowns.

Example 1) For each of the following systems of equations, using "**row reduction**", the system has been reduced so that the solution can be easily (**trust me!**) determined. State the solutions for each system.

(a)
$$\begin{array}{l} x + 2y - 6z = 5 \\ 4x + 5y - 21z = 5 \\ -3x + 3y + 17z = -2 \end{array}$$
which can be reduced to
$$\begin{array}{l} x + 2y - 6z = 5 \\ y - z = 5 \\ z = -4 \end{array}$$

(b)
$$\begin{array}{l} x - 2y + 4z = 2 \\ 2x - 3y + 5z = 3 \\ 3x - 4y + 6z = 7 \end{array}$$
which can be reduced to
$$\begin{array}{l} x + 2y - 3z = 2 \\ y - 3z = -1 \\ 0 = 3 \end{array}$$

(c)
$$\begin{array}{l} x + 2y + 3z = 3 \\ 2x + 3y + 8z = 4 \\ 3x + 2y + 17z = 1 \end{array}$$
which can be reduced to
$$\begin{array}{l} x + 2y + 3z = 3 \\ y - 2z = 2 \\ 0 = 0 \end{array}$$

Solution 1)(a) Unique solution. The three planes intersect in a single point: $z = -4$, $y = 1$, and $x = -21$

(b) Inconsistent solution. The three planes have no common point of intersection. This can happen when at least two of the planes are parallel but not coincident.

(c) Consistent (dependent) solution. This happens when at least two of the planes coincide or no two of the three are parallel but they have a common line of intersection. If all three planes coincide, you obtain only one non-trivial equation, that is one non-"0 = 0" equation.
Here, with "free" variable (or "parameter") z, we have $y = 2 + 2z$ and so
$x = 3 - 3z - 2(2 + 2z) = 3 - 3z - 4 - 4z = -1 - 7z$.

Four for you.

The following equations are the reduced forms of four linear systems in three variables. State the solution for each and whether it is dependent, inconsistent, and/or unique.

1) $x + y + z = 3$, $y - z = 5$, $z = 1$

2) $x + y + z = 3$, $y - z = 5$, $0 = 0$

3) $x + y + z = 3$, $0 = 0$, $0 = 0$

4) $x + y + z = 3$, $y - z = 5$, $0 = 1$

Answers 1) $z = 1$, $y = 6$, $x = -4$. The system has a unique solution.

2) $z \in \mathbb{R}$, $y = 5 + z$, $x = -2 - 2z$; dependent system with infinite solutions (one free variable)

3) $z \in \mathbb{R}$, $y \in \mathbb{R}$, $x = 3 - y - z$; dependent system with infinite solutions (two free variables).

4) No solution. The system is inconsistent.

Solving Quadratic Equations
Using the Quadratic Formula

Some quadratics factor and solve very easily, such as $x^2 - 4x + 3 = (x-3)(x-1) = 0$.

Others, such as $2x^2 - x - 5 = 0$, have "less pleasant" real roots. Still others, such as $2x^2 - x + 3 = 0$ have non-real roots. In these latter cases, the quadratic formula makes solving for x much simpler.

Remember the quadratic formula : if $ax^2 + bx + c = 0$, then $x = \dfrac{-b \pm \sqrt{b^2 - 4ac}}{2a}$

Example 1) Solve the equation $2x^2 - x - 5 = 0$ using the quadratic formula.

Solution $a = 2$, $b = -1$, $c = -5$ $\therefore$ $x = \dfrac{-(-1) \pm \sqrt{(-1)^2 - 4(2)(-5)}}{2(2)} = \dfrac{1 \pm \sqrt{41}}{4}$

The roots are $x = \dfrac{1 + \sqrt{41}}{4}$ and $x = \dfrac{1 - \sqrt{41}}{4}$.

Example 2) Solve the equation $2x^2 - x + 3 = 0$ using the quadratic formula.

Solution $a = 2$, $b = -1$, $c = 3$ $\therefore$ $x = \dfrac{-(-1) \pm \sqrt{(-1)^2 - 4(2)(3)}}{2(2)} = \dfrac{1 \pm \sqrt{-23}}{4} = \dfrac{1 \pm \sqrt{23}\,i}{4}$

The roots are $x = \dfrac{1 + i\sqrt{23}}{4}$ and $x = \dfrac{1 - i\sqrt{23}}{4}$.

Two for you.

Solve each of the following using the quadratic formula:

1) $3x^2 - 4x - 1 = 0$ 2) $5x^2 + 6x + 5 = 0$

Answers 1) $\dfrac{2+\sqrt{7}}{3}$, $\dfrac{2-\sqrt{7}}{3}$ 2) $\dfrac{-3+4i}{5}$, $\dfrac{-3-4i}{5}$

Factoring Quadratic Expressions Using the Quadratic Formula

Some quadratic expressions factor very easily, such as $x^2 - 4x + 3 = (x-3)(x-1)$. When we need to factor quadratic expressions whose corresponding equations have "unpleasant real roots, such as $2x^2 - x - 5 = 0$, or non-real roots, such as $2x^2 - x + 3 = 0$, we can use the quadratic formula to make factoring simple.

Remember the quadratic formula: if $ax^2 + bx + c = 0$, then $x = \dfrac{-b \pm \sqrt{b^2 - 4ac}}{2a}$

Here is a quadratic fact of life. Let $f(x) = ax^2 + bx + c$ have roots r_1 and r_2. Then we can factor $f(x) = ax^2 + bx + c = a(x - r_1)(x - r_2)$. **Please note well the "a" in the factored form.**

Example 1) Factor $f(x) = 2x^2 - x - 5$ using the quadratic formula.

Solution $a = 2$, $b = -1$, $c = -5$ $\therefore$ $x = \dfrac{-(-1) \pm \sqrt{(-1)^2 - 4(2)(-5)}}{2(2)} = \dfrac{1 \pm \sqrt{41}}{4}$

The roots are $r_1 = \dfrac{1 + \sqrt{41}}{4}$ and $r_2 = \dfrac{1 - \sqrt{41}}{4}$

$\therefore$ $f(x) \overset{a(x-r_1)(x-r_2)}{=} 2\left(x - \dfrac{1 + \sqrt{41}}{4}\right)\left(x - \dfrac{1 - \sqrt{41}}{4}\right) \overset{\text{if you prefer...}}{=} \left(2x - \dfrac{1 + \sqrt{41}}{2}\right)\left(x - \dfrac{1 - \sqrt{41}}{4}\right)$

Example 2) Factor the expression $g(x) = 2x^2 - x + 3$ using the quadratic formula.

Solution $a = 2$, $b = -1$, $c = 3$ $\therefore$ $x = \dfrac{-(-1) \pm \sqrt{(-1)^2 - 4(2)(3)}}{2(2)} = \dfrac{1 \pm \sqrt{-23}}{4} = \dfrac{1 \pm \sqrt{23}\, i}{4}$

The roots are $r_1 = \dfrac{1 + i\sqrt{23}}{4}$ and $r_2 = \dfrac{1 - i\sqrt{23}}{4}$

$\therefore$ $g(x) \overset{a(x-r_1)(x-r_2)}{=} 2\left(x - \dfrac{1 + i\sqrt{23}}{4}\right)\left(x - \dfrac{1 - i\sqrt{23}}{4}\right) \overset{\text{or if you prefer...}}{=} \left(2x - \dfrac{1 + i\sqrt{23}}{2}\right)\left(x - \dfrac{1 - i\sqrt{23}}{4}\right)$

Two for you.

Factor each of the following using the quadratic formula:

1) $3x^2 - 4x - 1$ 2) $5x^2 + 6x + 5$

Answers 1) $3\left(x - \dfrac{2+\sqrt{7}}{3}\right)\left(x - \dfrac{2-\sqrt{7}}{3}\right) \overset{\boxed{\text{or if you prefer...}}}{=} 3\left(3x - 2 - \sqrt{7}\right)\left(x - \dfrac{2-\sqrt{7}}{3}\right)$

2) $5\left(x - \dfrac{-3+4i}{5}\right)\left(x - \dfrac{-3-4i}{5}\right) \overset{\boxed{\text{better written as...}}}{=} \left(5x + 3 - 4i\right)\left(x + \dfrac{3+4i}{5}\right)$

Problems Involving the Sum and Product of the Roots of a Quadratic Equation

Remember the quadratic formula : if $ax^2 + bx + c = 0$, then $x = \dfrac{-b \pm \sqrt{b^2 - 4ac}}{2a}$

Adding the two roots: $\dfrac{-b + \sqrt{b^2 - 4ac}}{2a} + \dfrac{-b - \sqrt{b^2 - 4ac}}{2a} = \dfrac{-2b}{2a} = -\dfrac{b}{a}$

Multiplying the two roots: $\left(\dfrac{-b + \sqrt{b^2 - 4ac}}{2a}\right)\left(\dfrac{-b - \sqrt{b^2 - 4ac}}{2a}\right) = \dfrac{b^2 - (b^2 - 4ac)}{4a^2} = \dfrac{4ac}{4a^2} = \dfrac{c}{a}$

Example 1) Identify the sum and product of the roots of $3x^2 - 4x - 2 = 0$ **without solving for the roots**!

Solution $a = 3$, $b = -4$, $c = -2$ $\therefore$ $r_1 + r_2$ $\boxed{\text{Sum} = -\frac{b}{a}}$ $= -\dfrac{(-4)}{3} = \dfrac{4}{3}$ and $r_1 r_2$ $\boxed{\text{Product} = \frac{c}{a}}$ $= -\dfrac{2}{3}$

Example 2) The sum and product of the roots of a quadratic equation are $\dfrac{4}{3}$ and $-\dfrac{2}{3}$.

Find the quadratic equation.

Solution Rewrite $ax^2 + bx + c = 0$ as $x^2 + \dfrac{b}{a}x + \dfrac{c}{a} = 0 = x^2 - \left(-\dfrac{b}{a}\right)x + \dfrac{c}{a}$.

We are given $r_1 + r_2 = \dfrac{4}{3} = -\dfrac{b}{a}$ and $r_1 r_2 = -\dfrac{2}{3} = \dfrac{c}{a}$.

Therefore, the required quadratic equation is

$x^2 - \left(\dfrac{4}{3}\right)x - \dfrac{2}{3} = 0$ or $3x^2 - 4x - 2 = 0$.

Two for you.

1) Identify the sum and product of the roots of $\pi x^2 + ex - 1 = 0$ **without solving for the roots!**

2) The sum and product of the roots of a quadratic equation are $-\dfrac{2}{5}$ and 3. Find the quadratic equation.

Answers 1) Sum $= -\dfrac{e}{\pi}$, Product $= -\dfrac{1}{\pi}$

2) $x^2 + \dfrac{2}{5}x + 3 = 0$ or $5x^2 + 2x + 15 = 0$

The Graph of $y = a(x - b)^2 + c$

Given the parabola $y = a(x-b)^2 + c$, the vertex is (b,c) and the graph opens up if $a > 0$ and down if $a < 0$. The y intercept (where $x = 0$) is $ab^2 + c$.

Example 1) State the vertex and y intercept and draw the graph for each of the following: (a) $y = (x-2)^2 + 1$ (b) $y = -2(x+2)^2 + 4$

Solution (a) vertex: $(2,1)$; y intercept $= 5$ (b) vertex: $(-2,4)$; y intercept $= -4$

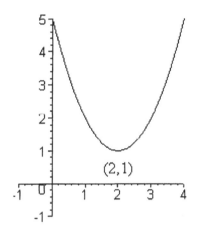
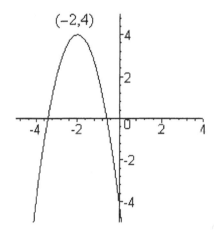

Example 2) Let $y = 1 - x - x^2$. Find the vertex and draw the graph.

Solution To find the vertex, **complete the square**.

| The coefficient of x is -1. Divide by 2 and square! | | We subtracted 1/4 so we add 1/4. |

$$y = -x - x^2 = -(x^2 - x) \quad = \quad -\left(x^2 - x + \frac{1}{4}\right) + \frac{1}{4} = -\left(x - \frac{1}{2}\right)^2 + \frac{1}{4}$$

The vertex is $\left(\dfrac{1}{2}, \dfrac{1}{4}\right)$

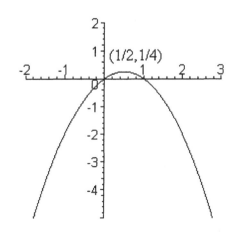

41

Two for you.

State the vertex, the y intercept, and whether the parabola opens up or down:

1) $y = 2(x+3)^2 - 5$ 2) $y = 1 - 8x - 2x^2$

Answers 1) vertex: $(-3,-5)$; y intercept $= 13$; up   2) vertex: $(-2,9)$; y intercept $= 1$; down

Completing the Square

Let's set this up so that we always have $a > 0$:

$$(x + a)^2 = x^2 + 2ax + a^2 \qquad (x - a)^2 = x^2 - 2ax + a^2$$

Note the sequence a, $2a$, and a^2. Completing the square is a piece of (quadratic) cake! All you do is

1) take the $2a$ term   **2) divide by 2 to get the a term**   **3) square to get the a^2 term.**

Example 1) Complete the square in the expression $4x^2 + 12x - 17$.

Solution $4x^2 + 12x - 17$

> Factor the 4 from the x^2 term and the x term BUT NOT THE CONSTANT!

$$= \qquad 4(x^2 + 3x) - 17$$

> Now, $2a = 3$, so $a = 3/2$ and $a^2 = 9/4$.
> By the way, note how we got "$3x$": $3x = 12x/4$
> Keep this in mind for the NEXT example.

$$= \qquad 4(x^2 + 3x + \frac{9}{4}) - 17 - 4(9/4) \qquad \boxed{\text{We added } 4\left(\frac{9}{4}\right) \text{ so we subtracted } 4\left(\frac{9}{4}\right).}$$

$$= \qquad 4\left(x + \frac{3}{2}\right)^2 - 26$$

Note: **Always** factor out the coefficient of x^2. Factor it from the x term as well but **NOT** the constant.

Example 2) Complete the square in the expression $5 - \frac{3}{2}x^2 + 2x$.

> Factor out $-\frac{3}{2}$. Note: $\frac{4}{3}x = \frac{2x}{\left(\frac{3}{2}\right)}$

Solution $5 - \frac{3}{2}x^2 + 2x \qquad = \qquad -\frac{3}{2}\left(x^2 - \frac{4}{3}x\right) + 5$

$$= -\frac{3}{2}\left(x^2 - \frac{4}{3}x + 4/9\right) + 5 + \frac{3}{2}\left(\frac{4}{9}\right) \quad \boxed{\begin{array}{l} \text{Note } 2a = -4/3, \text{ so } a = -2/3, \text{ and } a^2 = 4/9. \\[4pt] \text{Also, we } \textbf{SUBTRACTED } \frac{3}{2}\left(\frac{4}{9}\right) \text{ so,} \\[4pt] \text{to compensate, we } \textbf{ADD } \frac{3}{2}\left(\frac{4}{9}\right). \end{array}}$$

$$= -\frac{3}{2}\left(x - \frac{2}{3}\right)^2 + 5 + \frac{2}{3}$$

$$= -\frac{3}{2}\left(x - \frac{2}{3}\right)^2 + \frac{17}{3}$$

Two for you.

Complete the square for each of the following:

1) $3x^2 - 30x - 11$ 2) $x - x^2$

Answers 1) $3(x-5)^2 - 86$ 2) $\dfrac{1}{4} - \left(x - \dfrac{1}{2}\right)^2$

Solving Linear Inequalities

Now pay attention!

1) One of math's cardinal rules: **what you do to one side you do to the other!**

2) Not cardinal but important: when you multiply or divide an **inequality** by a **negative**, the direction of the inequality **reverses**.

For example, $-2 < 4$. Multiply both sides by -3 and you get $6 \boxed{\text{The direction is reversed!}} > -12$.

Example 1) Solve the inequality $4 - 2x < 5 + 8x$.

Solution $4 - 2x < 5 + 8x$. Bring the x terms to the left and the numbers to the right.

$-10x < 1$. Now divide both sides by -10: $x \boxed{\begin{array}{l}\text{We divided by a negative so}\\ \text{the inequality reverses!}\end{array}} > -\dfrac{1}{10}$

OR

$4 - 2x < 5 + 8x$. Bring the numbers to the left and the x terms to the right: $-1 < 10x$

Now divide both sides by 10: $-\dfrac{1}{10} \boxed{\begin{array}{l}\text{This time the inequality direction \textbf{DID NOT CHANGE}}\\ \text{because we divided by a \textbf{POSITIVE} number!}\end{array}} < x$

Example 2) Solve the inequality $3 \le 2x - 5 < 7$.

Solution $3 \le 2x - 5 < 7 \quad \boxed{\begin{array}{l}\text{Add 5 to each of the three}\\ \text{parts of the inequality...}\end{array}} \Leftrightarrow \quad 8 \le 2x < 12 \quad \boxed{\text{...and divide by 2.}} \Leftrightarrow \quad 4 \le x < 6$

Example 3) Solve the inequality $3 - 2x < 6 + 4x < 7$.

Solution This time we **MUST** solve two inequalities separately, because x appears more than once.

$3 - 2x < 6 + 4x$ and $6 + 4x < 7$ (We need to say "**AND**" because **both** inequalities must be satisfied!)

$-6x < 3 \qquad$ and $\qquad 4x < 1$

$x > -\dfrac{1}{2} \qquad$ and $\qquad x < \dfrac{1}{4} \qquad \therefore \quad -\dfrac{1}{2} < x < \dfrac{1}{4}$

45

Two for you.

Solve these inequalities:

1) $3 > 5 - 6x \geq 2$ 2) $3x + 7 < 4x + 5 < -x + 5$

Answers 1) $\dfrac{1}{3} < x \leq \dfrac{1}{2}$

2) No solution.

$(3x + 7 < 4x + 5 \Rightarrow x > 2$ while $4x + 5 < -x + 5 \Rightarrow x < 0$.

There are no numbers than are greater than 2 **AND** less than 0!)

Solving Quadratic Inequalities

Some of these inequalities factor and solve very easily. Some don't factor, which means there are no intercepts: the parabola is either always above or always below the x axis. Some factor if you first **complete the square** and then use **difference of squares**.

Example 1) Solve the inequality $x^2 - 3x - 4 > 0$.

Solution $x^2 - 3x - 4 > 0 \Leftrightarrow (x-4)(x+1) > 0$

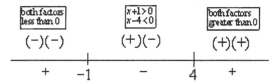

both factors less than 0	$x+1>0$ $x-4<0$	both factors greater than 0
$(-)(-)$	$(+)(-)$	$(+)(+)$

$$+ \quad -1 \quad - \quad 4 \quad +$$

The solution is $x \in (-\infty, -1) \cup (4, \infty)$.

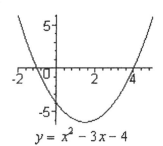

$y = x^2 - 3x - 4$

Example 2) Solve the inequality $x^2 - x + 1 < 0$.

Solution $x^2 - x + 1 < 0$

Completing the square
$(x+a)^2 = x^2 + 2ax + a^2$
Here, $a = -\dfrac{1}{2}$.

$$\Leftrightarrow \quad x^2 - x + \frac{1}{4} + 1 - \frac{1}{4} < 0 \Leftrightarrow \left(x - \frac{1}{2}\right)^2 + \frac{3}{4} < 0$$

Since $\left(x - \dfrac{1}{2}\right)^2 + \dfrac{3}{4} \geq \dfrac{3}{4} \overset{\text{Always!}}{>} 0$, there is no solution.

(Sometimes **BOB** – that is, the **Back Of** the **Book** – calls no solution "**the null set**" or "**the empty set**".)

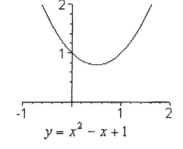

$y = x^2 - x + 1$

Example 3) Solve the inequality $x^2 - x - 1 < 0$.

Solution $x^2 - x - 1 < 0 \Leftrightarrow x^2 - x + \dfrac{1}{4} - 1 - \dfrac{1}{4} < 0 \Leftrightarrow \left(x - \dfrac{1}{2}\right)^2 - \dfrac{5}{4} < 0$

$a^2 - b^2 = (a-b)(a+b)$
$a = x - \dfrac{1}{2}$ $b = \dfrac{\sqrt{5}}{2}$

$$\Leftrightarrow \left(x - \frac{1}{2} - \frac{\sqrt{5}}{2}\right)\left(x - \frac{1}{2} + \frac{\sqrt{5}}{2}\right) < 0$$

Be careful with "–" when you make "2" a common denominator for each root.

$$\Leftrightarrow \left(x - \frac{1+\sqrt{5}}{2}\right)\left(x - \frac{1-\sqrt{5}}{2}\right) < 0$$

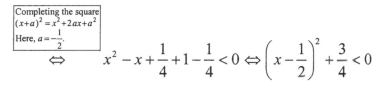

$(-)(-)$	$(+)(-)$	$(+)(+)$

$$+ \quad \frac{1-\sqrt{5}}{2} \quad - \quad \frac{1+\sqrt{5}}{2} \quad +$$

$$\therefore \quad x \in \left(\frac{1-\sqrt{5}}{2}, \frac{1+\sqrt{5}}{2}\right)$$

Two for you.

Solve the inequalities: 1) $x^2 + 7x + 12 > 0$ 2) $x^2 + 2x - 1 < 0$

Answers 1) $x \in (-\infty, -4) \cup (-3, \infty)$ 2) $x \in \left(-1 - \sqrt{2}, -1 + \sqrt{2}\right)$

Solving Inequalities with Two or More Factors

Here, you will have an inequality where

1) on one side you will have a **product** and/or **quotient** of factors in the form $x \pm a$

2) on the other side **you must have 0!**

Example 1) Solve the inequality $(x+3)x(x-4) < 0$.

Solution

$$
\begin{array}{ccccc}
(-)(-)(-) & (+)(-)(-) & (+)(+)(-) & (+)(+)(+) \\
\end{array}
$$

$$
\underset{-}{\rule{0pt}{0pt}} \quad \underset{-3}{|} \quad \underset{+}{\rule{0pt}{0pt}} \quad \underset{0}{|} \quad \underset{-}{\rule{0pt}{0pt}} \quad \underset{4}{|} \quad \underset{+}{\rule{0pt}{0pt}}
$$

$\therefore \ x \in (-\infty, -3) \cup (0, 4)$

Example 2) Solve the inequality $(x+3)^2 (x+1)^3 x^{1/3} (x-4) \leq 0$.

Solution This time, the factor $(x+3)$ **DOESN'T MATTER** because it is raised to an **EVEN** exponent: $(x+3)^2 \geq 0$ **ALWAYS**. I will include -3 on the number line just for emphasis! However, $x+1$ is raised to an **ODD** exponent and so it will change from $-$ to $+$ as x goes from less than -1 to greater than -1. Ditto for the $x^{1/3}$ term as x goes from less than 0 to greater than 0.

$$
\begin{array}{ccccc}
(-)(-)(-) & (-)(-)(-) & (+)(-)(-) & (+)(+)(-) & (+)(+)(+) \\
\end{array}
$$

$$
\underset{-}{\rule{0pt}{0pt}} \quad \underset{-3}{|} \quad \underset{-}{\rule{0pt}{0pt}} \quad \underset{-1}{|} \quad \underset{+}{\rule{0pt}{0pt}} \quad \underset{0}{|} \quad \underset{-}{\rule{0pt}{0pt}} \quad \underset{4}{|} \quad \underset{+}{\rule{0pt}{0pt}}
$$

The solution is $x \in (-\infty, -1] \cup [0, 4]$.

Example 3) Solve the inequality $\dfrac{(x+3)(x+1)x}{(x-4)} \leq 0$.

Solution Watch out for division by 0: we **can't** let x be 4.

$$
\begin{array}{ccccc}
(-)(-)(-)(-) & (+)(-)(-)(-) & (+)(+)(-)(-) & (+)(+)(+)(-) & (+)(+)(+)(+) \\
\end{array}
$$

$$
\underset{+}{\rule{0pt}{0pt}} \quad \underset{-3}{|} \quad \underset{-}{\rule{0pt}{0pt}} \quad \underset{-1}{|} \quad \underset{+}{\rule{0pt}{0pt}} \quad \underset{0}{|} \quad \underset{-}{\rule{0pt}{0pt}} \quad \underset{4}{|} \quad \underset{+}{\rule{0pt}{0pt}}
$$

$\therefore \ x \in [-3, -1] \cup [0, 4)$

Two for you.

Solve these inequalities:

1) $(2x-3)(4-x)(x-7)^3(x+1)^2 > 0$

$$\left(\text{Hint: write the inequality as } -2\left(x-\frac{3}{2}\right)(x-4)(x-7)^3(x+1)^2 > 0 \text{ and then} \right.$$

$$\left. \left(x-\frac{3}{2}\right)(x-4)(x-7)^3(x+1)^2 < 0. \right)$$

2) $\dfrac{(x+3)^2(x-1)}{(x+5)^{3/2}(x-2)(x-4)^3} \leq 0$ (Hint: $x > -5$; otherwise, $1/(x+5)^{3/2}$ is undefined.)

Answers 1) $x \in \left(-\infty, \dfrac{3}{2}\right) \cup (4,7)$ 2) $x \in (-5,1] \cup (2,4)$

Solving Rational Inequalities

In these questions, if you **cross multiply**, you need **SEPARATE CASES**.

Multiply by a " + " and the direction stays the same!
Multiply by a " − " and the direction reverses!

HERE IS AN EASIER WAY...

Example 1) Solve the inequality $\dfrac{1}{x+2} \le \dfrac{2}{3x+1}$.

Solution $\dfrac{1}{x+2} \le \dfrac{2}{3x+1}$ $\boxed{\text{Do not cross-mulitply!}}$ $\Leftrightarrow$ $\dfrac{1}{x+2} - \dfrac{2}{3x+1} \le 0$ $\boxed{\text{Get a common denominator.}}$ $\Leftrightarrow$ $\dfrac{3x+1-2(x+2)}{(x+2)(3x+1)} \le 0$

$\Leftrightarrow \dfrac{x-3}{3(x+2)\left(x+\dfrac{1}{3}\right)} \le 0$

$$\begin{array}{ccccccccc} (-)(-)(-) & & (+)(-)(-) & & (+)(+)(-) & & (+)(+)(+) \\ \hline & -2 & & -\dfrac{1}{3} & & 3 & \\ - & & + & & - & & + \end{array}$$

The solution is $x \in (-\infty, -2) \cup \left(-\dfrac{1}{3}, 3\right]$.

Example 2) Solve the inequality $\dfrac{2x-1}{3x+1} \ge \dfrac{x+2}{x-2}$.

Solution $\dfrac{2x-1}{3x+1} \ge \dfrac{x+2}{x-2}$ $\boxed{\text{Don't cross multiply!}}$ $\Leftrightarrow$ $\dfrac{2x-1}{3x+1} - \dfrac{x+2}{x-2} \ge 0$ $\boxed{\text{common denominator}}$ $\Leftrightarrow$ $\dfrac{(2x-1)(x-2)-(x+2)(3x+1)}{(3x+1)(x-2)} \ge 0$

$\boxed{\text{Expand...}}$ $\Leftrightarrow \dfrac{2x^2-5x+2-(3x^2+7x+2)}{(3x+1)(x-2)} \ge 0$ $\boxed{\text{...and simplify.}}$ $\Leftrightarrow \dfrac{-x^2-12x}{(3x+1)(x-2)} \ge 0$ $\boxed{\text{Multiply both sides by }-1.}$ $\Leftrightarrow \dfrac{x^2+12x}{(3x+1)(x-2)} \le 0$

$\Leftrightarrow \dfrac{x(x+12)}{3\left(x+\dfrac{1}{3}\right)(x-2)} \le 0$

$$\begin{array}{ccccccccccc} (-)(-)(-)(-) & & (+)(-)(-)(-) & & (+)(+)(-)(-) & & (+)(+)(+)(-) & & (+)(+)(+)(+) \\ \hline & -12 & & -\dfrac{1}{3} & & 0 & & 2 & \\ + & & - & & + & & - & & + \end{array}$$

Therefore, $x \in \left[-12, -\dfrac{1}{3}\right) \cup [0, 2)$.

51

Two for you.

Solve: 1) $\dfrac{5}{x+7} \geq \dfrac{2}{x-5}$ 2) $\dfrac{x-1}{x+1} \leq \dfrac{3x-1}{3x+1}$

Answers 1) $(-7,5) \cup [13,\infty)$ 2) $\left(-1,-\dfrac{1}{3}\right) \cup [0,\infty)$

The Basics of Absolute Value

Remember, absolute value is **ALWAYS POSITIVE!**

$|3| = 3$, $\quad |-3| = 3 \overset{\boxed{\begin{array}{l}\text{This step is}\\ \text{THE KEY STEP}\\ \text{for understanding}\\ \text{absolute value!}\end{array}}}{=} -(-3)$ and so $\quad |x| = \begin{cases} -x, & if \ x < 0 \\ x, & if \ x \geq 0 \end{cases}$

Here is the part that people find so confusing.

WHY PUT "–" IN FRONT OF THE x?

Why not just make it positive as we did with -3?

BECAUSE ONE OF THE "–" SIGNS IS **INSIDE** THE x, **so you can't get rid of it!**

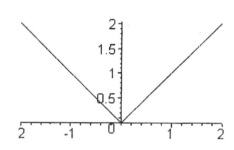

$f(x) = |x| = \begin{cases} -x, & if \ x < 0 \\ x, & if \ x \geq 0 \end{cases}$

Domain $= \mathbb{R}$, Range $= [0, \infty)$

Example 1) Write (a) $|x-1|$ (b) $|x+1|$ without using the absolute value notation.

Solution (a) $|x-1| = \begin{cases} -(x-1), & \text{if } x < 1 \\ x-1, & \text{if } x \geq 1 \end{cases}$ (b) $|x+1| = \begin{cases} -(x+1), & \text{if } x < -1 \\ x+1, & \text{if } x \geq -1 \end{cases}$

Example 2) Graph: (a) $y = |x-1|$ (b) $y = |x+1|$.

Solution (a) $y = |x-1|$ (b) $y = |x+1|$

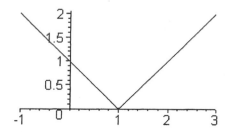 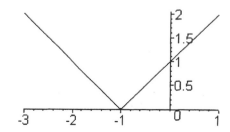

Two for you.

1) Write $|2x - 3|$ without absolute value signs.

2) Graph $y = |2x - 3|$.

Answers 1) $|2x - 3| = \begin{cases} -(2x - 3), & \text{if } x < \dfrac{3}{2} \\[2mm] (2x - 3), & \text{if } x \geq \dfrac{3}{2} \end{cases}$ 2)

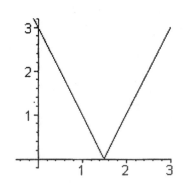

Solving Absolute Value Equations

Remember, absolute value is **ALWAYS POSITIVE!**

Example 1) Solve the following absolute value equations.

(a) $|x| = 4$ (b) $|x| = -3$ (c) $|2x - 5| = 7$ (d) $|x + 5| = 2x - 3$ (e) $|2x + 1| = |x - 7|$

Solution (a) $x = 4$ or $x = -4$

(b) No solution since $|x|$ is positive.

(c) Either $2x - 5 = 7$ in which case $2x = 12$ and $x = 6$

or

$2x - 5 = -7$ in which case $2x = -2$ and $x = -1$.

(Remember, x can be negative; it is the absolute value of x that must be positive.)

(d) Easiest method here is to use cases.

Case 1) $x + 5 \geq 0$ so $x \geq -5$. In this case, $|x + 5| = x + 5$. So we solve:
$x + 5 = 2x - 3 \Leftrightarrow -x = -8 \Leftrightarrow x = 8$. Since $8 \geq -5$, $x = 8$ is a solution.

Case 2) $x + 5 < 0$ so $x < -5$. In this case, $|x + 5| = -(x + 5)$. So we solve:

$-(x + 5) = 2x - 3 \Leftrightarrow -x - 5 = 2x - 3 \Leftrightarrow -8 \Leftrightarrow -3x = 2 \Leftrightarrow x = -\dfrac{2}{3}$

Since $-\dfrac{2}{3}$ **IS NOT LESS THAN** -5, $x = -\dfrac{2}{3}$ is **NOT** a solution.

(e) Since $|a| = |b| \Leftrightarrow a^2 = b^2$, the easiest method here is to **square both sides**.

$|2x + 1| = |x - 7| \Leftrightarrow (2x + 1)^2 = (x - 7)^2 \Leftrightarrow 4x^2 + 4x + 1 = x^2 - 14x + 49$

$\Leftrightarrow 3x^2 + 18x - 48 = 0 \Leftrightarrow (3x - 6)(x + 8) = 0 \Leftrightarrow 3(x - 2)(x + 8) = 0 \Leftrightarrow x = 2$ or $x = -8$

We **DON'T** have to check our answers in this example because of "$\Leftrightarrow$"! Not only does each step follow from the previous step, each step is **REVERSIBLE**!
In example (d), if you use the method of squaring both sides, **YOU DO HAVE TO CHECK BECAUSE THE SQUARING IN THAT EXAMPLE IS NOT REVERSIBLE!**

Two for you.

1) Solve: (a) $|x^2| = 100$ (b) $|2x+1| = |3x-2|$ (c) $|3x+2| = |x-6|$

Answers 1)(a) $x = 10$ or $x = -10$ (b) $x = 3$ or $x = \dfrac{1}{5}$ (c) $x = -4$ or $x = 1$

Solving Easy Absolute Value Inequalities

Keep in mind that for $a > 0$: $\boxed{|x| < a \Leftrightarrow -a < x < a}$ and $\boxed{|x| > a \Leftrightarrow x < -a \text{ OR } x > a}$

Example 1) Solve the following absolute value inequalities.

(a) $|x| < 1$ (b) $|x| \geq 4$ (c) $|x| < -1$ (d) $|x| > -2$

Solution (a) $|x| < 1 \Leftrightarrow -1 < x < 1$

(b) $|x| \geq 4 \Leftrightarrow x \leq -4 \text{ or } x \geq 4$

(c) $|x| < -1$ is **ALWAYS FALSE**, since $|x| \geq 0$.

(d) $|x| > -2 \Leftrightarrow x \in (-\infty, \infty)$, that is, $|x| > -2$ is **ALWAYS TRUE!**

Example 2) Solve: a) $|2x - 4| < 6$ (b) $|x - 3| \geq 7$

Solution (a) $|2x - 4| < 6 \Leftrightarrow -6 < 2x - 4 < 6 \Leftrightarrow -2 < 2x < 10 \Leftrightarrow -1 < x < 5$, that is, $x \in (-1, 5)$

(b) $|x - 3| \geq 7 \Leftrightarrow x - 3 \leq -7 \text{ or } x - 3 \geq 7 \Leftrightarrow x \leq -4 \text{ or } x \geq 10$, that is, $x \in (-\infty, -4] \cup [10, \infty)$

Two for you.
1) Solve: (a) $|x| \le 0.1$ (b) $|x| > 3.2$
2) Solve: (a) $|x+3| \le 2$ (b) $1 < |3x-5|$

Answers 1)(a) $[-0.1, 0.1]$ (b) $(-\infty, -3.2) \cup (3.2, \infty)$

2)(a) $[-5, -1]$ (b) $\left(-\infty, \dfrac{4}{3}\right) \cup (2, \infty)$

Solving Less Easy Absolute Value Inequalities

There are two basic methods to use when solving more complicated absolute value inequalities (and equations as well): "**cases**" or "**squaring both sides**". The key difference:

$|a| < b$ $\boxed{\text{The implication goes in only one direction!}}$ $\Rightarrow$ $a^2 < b^2$. Note that here, since $|a| \geq 0$, b **MUST BE POSITIVE!**

When you square, you **could** introduce solutions for $a^2 < b^2$ that don't work for $|a| < b$.

That's why you use cases in this type of problem.

However, $|a| < |b|$ $\boxed{\text{The implication goes in both directions!}}$ $\Leftrightarrow$ $a^2 < b^2$. Squaring both sides is faster than cases and solutions work both ways! **BUT...**

BE CAREFUL! $a < |b| \;\not\Rightarrow\; a^2 < b^2$: a^2 might be less than b^2, **but doesn't have to be!**

Example 1) Solve $|x + 3| < 2x$.

Solution Use cases since squaring both sides can lead to "extraneous" solutions.

Case 1) $x + 3 \geq 0$ so that $x \geq -3$. In this case, $|x+3| = x + 3$

The inequality becomes $x + 3 < 2x$ and so $3 < x$. The solution in this case is $x > 3$.

Case 2) $x + 3 < 0$ so that $x < -3$. In this case, $|x+3| = -(x+3)$

The inequality becomes $-x - 3 < 2x$. Therefore, $-3 < 3x$ and so $1 < x$.

Since $x > 1$ and $x < -3$ are incompatible, there are no solutions in this case.

Combining the two cases, the solution of $|x+3| < 2x$ is $x \in (3, \infty)$.

Example 2) Solve $|x + 4| < |2x - 6|$.

Solution In this example, squaring both sides is the best method.

(Note that there would be four separate cases, if we used the case method!)

$|x+4| < |2x-6| \Leftrightarrow x^2 + 8x + 16 < 4x^2 - 24x + 36 \Leftrightarrow 0 < 3x^2 - 32x + 20$

$\boxed{\text{Personal preference: I prefer the expression on the left and 0 on the right}}$ $\Leftrightarrow$ $3x^2 - 32x + 20 > 0 \Leftrightarrow (3x - 2)(x - 10) > 0 \Leftrightarrow \left(x - \dfrac{2}{3}\right)(x - 10) > 0$

$$
\begin{array}{ccccc}
(-)(-) & & (+)(-) & & (+)(+) \\
\end{array}
\qquad \therefore\; x \in \left(-\infty, \dfrac{2}{3}\right) \cup (10, \infty)
$$

$$
\underbrace{\qquad\qquad}_{+} \quad \dfrac{2}{3} \quad \underbrace{\qquad\qquad}_{-} \quad 10 \quad \underbrace{\qquad\qquad}_{+}
$$

Two for you.

Solve: 1) $5x - 8 \geq |4 - x|$ (Hint: use $|4 - x| = |x - 4|$.)   2) $|3x - 2| \leq |2x - 3$

Answers 1) $[2, \infty)$ 2) $(-1, 1)$

The Basics of $\sqrt{}$ and the Reason $\sqrt{x^2} = |x|$

Example 1) State the domain and range of the function $y = \sqrt{x}$ and draw the graph.

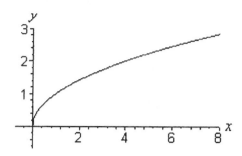

Solution The domain and range are both $[0, \infty)$.

Example 2) Solve: (a) $x = \sqrt{9}$ (b) $x = -\sqrt{9}$ (c) $x^2 = 9$

Solution (a) $x = \sqrt{9} = 3$ (b) $x = -\sqrt{9} = -3$ (c) $x^2 = 9 \implies x = 3$ or $x = -3$

Example 3) Evaluate $\sqrt{(3)^2}$, $\sqrt{(-3)^2}$, $|3|$, and $|-3|$. What does this tell you about $\sqrt{x^2}$?

Solution $\sqrt{(3)^2} = \sqrt{9} = 3$, $\sqrt{(-3)^2} = \sqrt{9} = 3$, $|3| = 3$,   and $|-3| = 3$.

$\therefore$ $\sqrt{x^2} = |x|$, for both $x > 0$ and $x < 0$. Put another way:

$$\sqrt{x^2} = \begin{cases} -x, \text{ if } x < 0 \\ x, \text{ if } x \geq 0 \end{cases}.$$

For example:

when $x = 3$, we have $\sqrt{(3)^2} \overset{\boxed{\begin{array}{c} x > 0 \text{ and so} \\ \sqrt{x^2} = x \end{array}}}{=} 3 = |3|$

when $x = -3$, we have $\sqrt{(-3)^2} = \sqrt{9} = 3 \overset{\boxed{\begin{array}{c} x < 0 \text{ and so} \\ \sqrt{x^2} = -x \end{array}}}{=} -(-3) = |-3|$

Two for you.

1) State the domain and range of $f(x) = \sqrt{x+1} + 1$ and draw the graph.

2) Write $y = \sqrt{(x-4)^2}$ using first absolute value and then a "branch" definition.

Answers 1) Domain $= [-1, \infty)$, Range $= [1, \infty)$

2) $y = |x - 4| = \begin{cases} -(x-4), \text{ if } x < 4 \\ x - 4, \text{ if } x \geq 4 \end{cases}$

$= \begin{cases} 4 - x, \text{ if } x < 4 \\ x - 4, \text{ if } x \geq 4 \end{cases}$

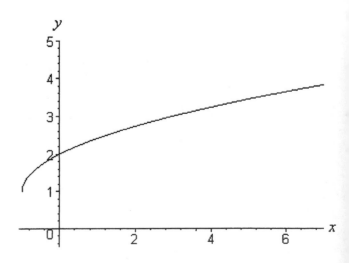

Solving Equations Involving Square Roots

Here, as with the more complicated absolute value questions, squaring both sides works.
BUT while $x = y \Rightarrow x^2 = y^2$, the converse is false: $x^2 = y^2 \not\Rightarrow x = y$. For example,
$(4)^2 = (-4)^2$ but $4 \neq -4$! When we square both sides, we can introduce "extraneous" roots.
So, we must take our possible solutions and check them in the original equations!

Example 1) Solve $\sqrt{x+5} = x - 7$.

Solution $\sqrt{x+5} = x - 7 \Rightarrow x + 5 = x^2 - 14x + 49 \Rightarrow x^2 - 15x + 44 = 0$
$\Rightarrow (x-11)(x-4) = 0 \Rightarrow x = 11$ or $x = 4$.

Check $x = 11$ in the original equation: Left Side $= \sqrt{11+5} = \sqrt{16} = 4$ Right side $= 11 - 7 = 4$

Check $x = 4$ in the original equation: Left Side $= \sqrt{4+5} = \sqrt{9} = 3$ Right side $= 4 - 7 = -3$
Therefore, the only solution is $x = 11$.

Example 2) Solve $\sqrt{2x-7} - \sqrt{x-4} = 1$

Solution $\sqrt{2x-7} - \sqrt{x-4} = 1$

| Isolate one of the square roots. |
$\Rightarrow \qquad \sqrt{2x-7} = \sqrt{x-4} + 1$

| Square both sides. |
$\Rightarrow \qquad 2x - 7 = x - 4 + 2\sqrt{x-4} + 1$

| Isolate the remaining square root. |
$\Rightarrow \qquad x - 4 = 2\sqrt{x-4}$

| Square both sides. |
$\Rightarrow \qquad x^2 - 8x + 16 = 4x - 16 \Rightarrow x^2 - 12x + 32 = 0 \Rightarrow (x-4)(x-8) = 0 \Rightarrow x = 4$ or $x = 8$

Check $x = 4$ in the original equation: Left Side $= \sqrt{2(4)-7} - \sqrt{4-4} = 1$ Right side $= 1$

Check $x = 8$ in the original equation: Left Side $= \sqrt{2(8)-7} - \sqrt{8-4} = 1$ Right side $= 1$
Therefore, both 4 and 8 are solutions.

Two for you.

1) Solve $\sqrt{2x-7}=x-3$.

2) Solve $\sqrt{x-3}-\sqrt{2x+1}=-2$ (Hint: first isolate the $\sqrt{2x+1}$ term.)

Answers 1) 4   2) 4, 12

Rationalizing Denominators that Have $\sqrt{}$

Often you run into problems where there is either a single term with a square root in the denominator or a binomial with one or two square roots. In the first case, a simple $\dfrac{\sqrt{}}{\sqrt{}}$ solves the problem. In the second, **DIFFERENCE OF SQUARES** comes to the rescue.

Example 1) Rationalize the denominators: (a) $\dfrac{3}{\sqrt{2}}$ (b) $\dfrac{\sqrt{7}}{2\sqrt{21}}$ (c) $\dfrac{xy}{\sqrt{2x}}$

Solution (a) $\dfrac{3}{\sqrt{2}} = \dfrac{3}{\sqrt{2}}\dfrac{\sqrt{2}}{\sqrt{2}} = \dfrac{3\sqrt{2}}{2}$

(b) $\dfrac{\sqrt{7}}{2\sqrt{21}} \overset{\boxed{\text{Only the }\sqrt{21}\text{ is important!}}}{=} \dfrac{\sqrt{7}}{2\sqrt{21}}\dfrac{\sqrt{21}}{\sqrt{21}} \overset{\boxed{\sqrt{7}\sqrt{21}=\sqrt{7}\sqrt{7}\sqrt{3}}}{=} \dfrac{7\sqrt{3}}{2(21)} = \dfrac{\sqrt{3}}{6}$

(c) $\dfrac{xy}{\sqrt{2x}} = \dfrac{xy}{\sqrt{2x}}\dfrac{\sqrt{2x}}{\sqrt{2x}} = \dfrac{\sqrt{2x}\,xy}{2x} = \dfrac{y\sqrt{2x}}{2}$

Example 2) Rationalize the denominators: (a) $-\dfrac{1}{\sqrt{x}-3}$ (b) $\dfrac{x}{\sqrt{x}+y}$ (c) $\dfrac{x}{\sqrt{2x+1}-3\sqrt{x-3}}$

Solution (a) $\dfrac{1}{\sqrt{x}-3} \overset{\boxed{\begin{array}{l}(a-b)(a+b)=a^2-b^2\\ \text{Here, }a=\sqrt{x}\text{ and }b=3.\end{array}}}{=} \dfrac{1}{\sqrt{x}-3}\left(\dfrac{\sqrt{x}+3}{\sqrt{x}+3}\right) = \dfrac{\sqrt{x}+3}{x-9}$

(b) $\dfrac{x}{\sqrt{x}+y} = \dfrac{x}{\sqrt{x}+y}\left(\dfrac{\sqrt{x}-y}{\sqrt{x}-y}\right) = \dfrac{x\sqrt{x}-xy}{x-y^2}$

(c) $\dfrac{x}{\sqrt{2x+1}-3\sqrt{x-3}} \overset{\boxed{\begin{array}{l}a=\sqrt{2x+1}\text{ and}\\ b=3\sqrt{x-3}\end{array}}}{=} \left(\dfrac{x}{\sqrt{2x+1}-3\sqrt{x-3}}\right)\left(\dfrac{\sqrt{2x+1}+3\sqrt{x-3}}{\sqrt{2x+1}+3\sqrt{x-3}}\right)$

$= \dfrac{x\left(\sqrt{2x+1}+3\sqrt{x-3}\right)}{2x+1-9(x-3)} = \dfrac{x\left(\sqrt{2x+1}+3\sqrt{x-3}\right)}{-7x+28}$

Two for you.

Rationalize the denominators: 1)(a) $\dfrac{\sqrt{3}}{\sqrt{45}}$ (b) $\dfrac{x-1}{\sqrt{x+4}}$ 2) $\dfrac{1}{\sqrt{x+h}-\sqrt{x}}$

Answers 1)(a) $\dfrac{\sqrt{15}}{15}$ (b) $\dfrac{(x-1)\sqrt{x+4}}{x+4}$ 2) $\dfrac{\sqrt{x+h}+\sqrt{x}}{h}$

Graphs of Basic Quadratic Relations

A quadratic relation in variables x and y is a parabola, circle, ellipse, or hyperbola.

Example 1) Draw the graphs of the following relations.

(a) Parabola: $y = x^2$ (b) Circle: $x^2 + y^2 = 1$

(c) Ellipse: $\dfrac{x^2}{9} + \dfrac{y^2}{25} = 1$ (d) Hyperbola: $x^2 - y^2 = 1$

Solution

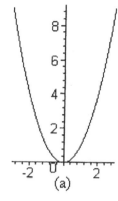

(a)

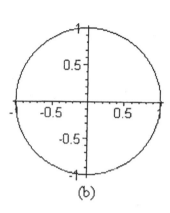

(b)

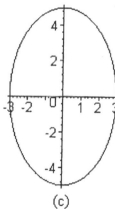

(c)

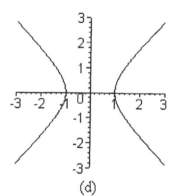

(d)

One for you.

1) Graph each of the following quadratic relations and identify it as a parabola, circle, ellipse, or hyperbola.

1)(a) $y^2 - x^2 = 4$ (b) $y = 3 - 2x^2$ (c) $x^2 + y^2 = 9$ (d) $\dfrac{x^2}{4} + y^2 = 1$

Answer

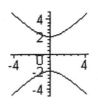 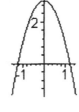 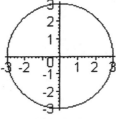 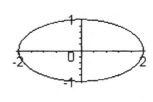

1)(a) hyperbola   (b) parabola   (c) circle   (d) ellipse

Basic $y = x^n$ Graphs, where $n \in \mathbb{N}$ and Why Even and Odd Functions Are Called Even and Odd Functions

Graphs with equations of the form $y = x^n$, where $n \in \mathbb{N}$, come up so often that they deserve a special page. This is it!

Example 1) Draw the graphs of $y = x^3$ and $y = x^5$ on the same axes.

Solution

For **odd** natural numbers n:

1) $\lim\limits_{x \to \pm\infty} x^n = \pm\infty$ and the expression approaches $\pm\infty$ faster as n increases.

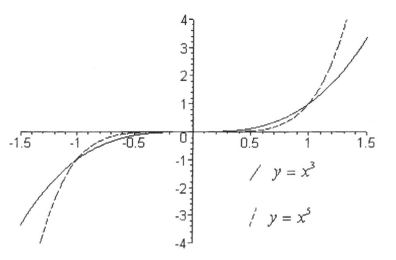

2) The graphs are symmetric in the **origin**, that is, $(-x)^n = -x^n$. This is the reason functions that satisfy $f(-x) = -f(x)$ are called **ODD** functions!

Example 2) Draw the graphs of $y = x^2$ and $y = x^4$ on the same axes.

Solution

For **even** natural numbers n:

1) $\lim\limits_{x \to \pm\infty} x^n = \infty$ and the expression approaches ∞ faster as n increases.

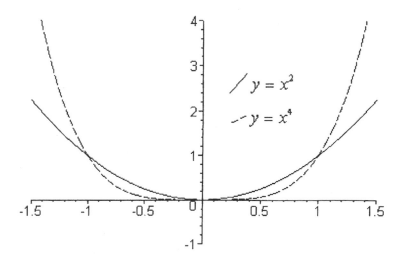

2) The graphs are symmetric in the y axis, that is, $(-x)^n = x^n$. This is the reason functions that satisfy $f(-x) = f(x)$ are called **EVEN** functions!

69

Two for you.

1) As $x \to \pm\infty$, which of these functions approach ∞ faster: $y = x^4$ or $y = x^6$?

2) As $x \to \pm\infty$, which of these functions approach $\pm\infty$ faster: $y = x^{11}$ or $y = x^9$?

Answers 1) $y = x^6$ 2) $y = x^{11}$

Basic $y = \dfrac{1}{x^n} = x^{-n}$ Graphs, where $n \in \mathbb{N}$

Graphs with equations of the form $y = \dfrac{1}{x^n}$, where $n \in \mathbb{N}$, come up so often that they deserve a special page. This is it!

Example 1) Draw the graphs of $y = \dfrac{1}{x}$ and $y = \dfrac{1}{x^3}$ on the same axes.

Solution

For **odd** natural numbers n:

1) $\displaystyle\lim_{x \to \pm\infty} \dfrac{1}{x^n} = 0$ and the expression approaches 0 faster as n increases.

2) $\displaystyle\lim_{x \to 0} \dfrac{1}{x^n} = \begin{cases} -\infty, & \text{if } x \to 0^- \\ \infty, & \text{if } x \to 0^+ \end{cases}$ and approaches $\pm\infty$ faster as n increases.

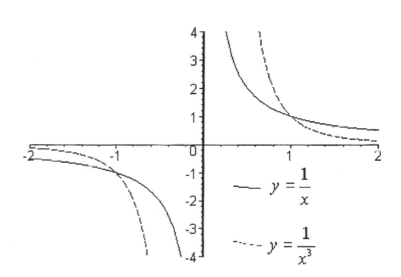

Example 2) Draw the graphs of $y = \dfrac{1}{x^2}$ and $y = \dfrac{1}{x^4}$ on the same axes.

Solution

For **even** natural numbers n:

1) $\displaystyle\lim_{x \to \pm\infty} \dfrac{1}{x^n} = 0$ and the expression approaches 0 faster as n increases.

2) $\displaystyle\lim_{x \to 0} \dfrac{1}{x^n} = \infty$ and approaches ∞ faster as n increases.

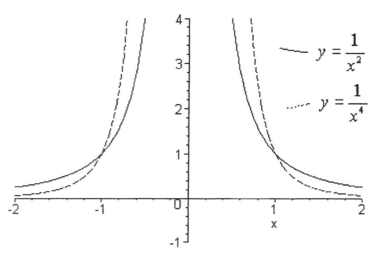

71

Two for you.

1) As $x \to \pm\infty$, which of these functions approach 0 faster: $y = x^{-4}$ or $y = x^{-5}$?

2) As $x \to 0^-$, which of these functions approach $-\infty$ faster: $y = x^{-11}$ or $y = x^{-9}$?

Answers 1) $y = x^{-5}$ 2) $y = x^{-11}$

Basic $y = x^{1/n}$ Graphs, where $n \in \mathbb{N}$

Graphs with equations of the form $y = x^{1/n}$, where $n \in \mathbb{N}$, come up so often that they deserve a special page. This is it!

Example 1) Draw the graphs of $y = x^{1/3}$ and $y = x^{1/5}$ on the same axes.

Solution

For **odd** natural numbers n:

1) $\lim\limits_{x \to \pm\infty} x^{1/n} = \pm\infty$ and the expression approaches $\pm\infty$ **more slowly** as n increases.

2) $\lim\limits_{x \to 0} x^{1/n} = 0$ and approaches 0 **more slowly** as n increases.

$$\left(\text{For example, } \left(\frac{1}{64}\right)^{1/3} = \frac{1}{4} < \left(\frac{1}{64}\right)^{1/6} = \frac{1}{2}. \right)$$

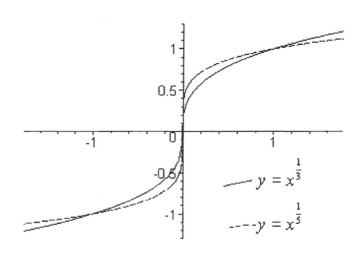

$$\underline{} \ y = x^{\frac{1}{3}}$$

$$\text{-----} \ y = x^{\frac{1}{5}}$$

Example 2) Draw the graphs of $y = x^{1/2}$ and $y = x^{1/4}$ on the same axes.

Solution For **even** natural numbers n:

1) The domain is $[0, \infty)$.

2) $\lim\limits_{x \to \infty} x^{1/n} = \infty$ and the expression approaches ∞ **more slowly** as n increases.

3) $\lim\limits_{x \to 0^+} x^{1/n} = 0$ and approaches 0 **more slowly** as n increases.

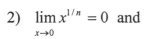

$$\underline{} \ y = x^{\frac{1}{2}}$$

$$\text{----} \ y = x^{\frac{1}{4}}$$

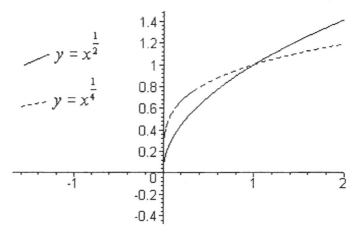

73

Two for you.

1) As $x \to \infty$, which of these functions approach ∞ fastest: $y = x^{1/4}$, $y = x^{1/5}$, or $y = x^{1/6}$?

2) As $x \to 0^-$, which of these functions approach 0 faster: $y = x^{1/11}$ or $y = x^{1/9}$?

Answers 1) $y = x^{1/4}$ 2) $y = x^{1/9}$

Shifting or Rescaling a Given Graph ($a > 0$)
$$f(x+a), \ f(x-a), \ f(ax), \ af(x), \ f(x)+a, \ f(x)-a$$

No matter what function $y = f(x)$ we begin with, the effect of each operation is **ALWAYS** the same.

Example 1) Let $y = f(x) = x^2$. Graph and describe the following functions relative to $f(x)$.

(a) $y = f(x+2) = (x+2)^2$

(b) $y = f(x-2) = (x-2)^2$

(c) $y = f(2x) = (2x)^2 = 4x^2$

(d) $y = 2f(x) = 2x^2$

(e) $y = f(x) + 2 = x^2 + 2$

(f) $y = f(x) - 2 = x^2 - 2$

Solution

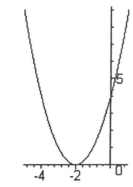

(a) Shifts $f(x)$ 2 units left.

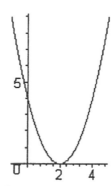

(b) Shifts $f(x)$ 2 units right.

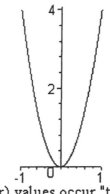

(c) $f(x)$ values occur "twice as fast

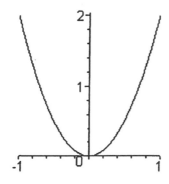

(d) Doubles the $f(x)$ values.

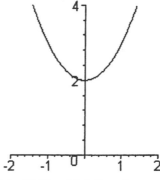

(e) Shifts $f(x)$ 2 units up.

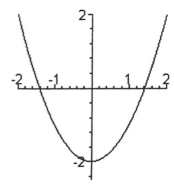

(f) Shifts $f(x)$ 2 units down.

Note: I had to choose from one of two ways to present these graphs. First: keep the scaling the same on all. This would clearly illustrate the changing shapes in (c) and (d). However, it would make it more difficult to see how the translations in (a), (b), (e), and (f) change the original graph, since I would have had to take x from –4 to 4. Second: adjust the scaling while keeping the graphs the same size. This works well for translations but not so well for shape. I decided rescaling was the better solution overall.

One for you.

To the right is the graph of $y = f(x)$ but **I WON'T** tell you the actual

function. Sketch (a) $y = f(x+0.5)$ (b) $y = f(x-0.5)$

(c) $y = f(0.5x)$ (d) $y = 0.5f(x)$

Answer

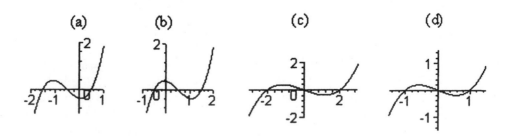

Tests for Symmetry

Symmetry in the y **axis**: both (x, y) and $(-x, y)$ are on the graph.

Replace x with $-x$ and see if you obtain the same y value.

Symmetry in the x **axis**: both (x, y) and $(x, -y)$ are on the graph.

Replace y with $-y$ and see if you obtain the same x value.

Symmetry in the origin: both (x, y) and $(-x, -y)$ are on the graph.

Replace both x with $-x$ and y with $-y$ and see if you obtain the same x and y values.

Example 1) Test for symmetry in the following relations.

(a) $y = x^2$ (b) $y = x^3 + x$ (c) $x = \cos(y)$ (d) $x^2 + y^2 = 25$

Solution (a) y axis (replace x with $-x$): $(-x)^2 = x^2$ and so there **is symmetry in the** y **axis**.

x axis (replace y with $-y$): $-y = -x^2 \neq x^2$ for $x \neq 0$ so there **is no symmetry in the** x **axis**

There is **no symmetry in the origin! WHY? BECAUSE...**

...you can have NO symmetry, ONE kind of symmetry, or ALL THREE.
But you <u>CANNOT</u> have exactly TWO KINDS OF SYMMETRY!

(b) y axis: $(-x)^3 + (-x) = -x^3 - x \neq x^3 + x$, for $x \neq 0$. NO!

x axis: $-y = -(x^3 + x) \neq x^3 + x$. NO!

Origin: $(-x)^3 + (-x) = -x^3 - x = -(x^3 + x) = -y$, and so $x^3 + x = y$. YES!

(c) y axis: $-x = -\cos y$. NO!

x axis: $\cos(-y) = \cos y = x$. YES!

Origin: NO!

(d) y axis: $(-x)^2 + y^2 = x^2 + y^2 = 25$. YES!

x axis: $x^2 + (-y)^2 = x^2 + y^2 = 25$. YES!

Origin: YES!

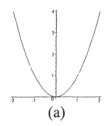
(a)

(b)

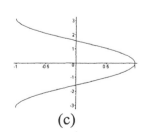
(c)

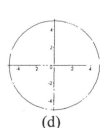
(d)

Two for you.

Discuss the symmetry for the following relations: 1) $y = x\, e^{|x|}$ 2) $x^3 + y^3 = x$

Answers 1) origin   2) origin

Graphing Polynomials without Calculus

Calculus tells us exactly where to find maximum and minimum points, the subtle changes at inflection points, and more. But with a little experience, we can tell a lot about the graph of a polynomial just from its equation.

Example 1) Determine the shape of the polynomial

$P(x) = 3x^5 - x^4 + 2x - 5$, as x approaches ∞ and $-\infty$.

Solution $P(x)$ does what its "leading term", "$3x^5$",

tells it to do when x is **BIG** in magnitude.

So, no matter what is happening "in the middle",

      this is the shape of $P(x)$ as $x \to \pm\infty$: ↗

The Shape in here
—— to be determined ——
using calculus!

If the polynomial is factored, we know its roots ($x - r$ is a factor $\Leftrightarrow$ $x = r$ is a root), and how it "behaves near the roots." If the exponent on the factor is **even**, the graph **doesn't change sign** as it passes through 0, that is, as it crosses the x axis at the root r. If the exponent is **odd**, **it does change sign**.

Example 2) Sketch the graph of $P(x) = x^2(x-1)^3$.

Solution The only roots are 0 and 1. From the exponents on the factors, we know $P(x) \le 0$ for $x \le 1$ and $P(x) \ge 0$ for $x \ge 1$ (the sign doesn't change at $x = 0$!). The leading term of $P(x)$, when expanded, is x^5. So, except for the "subtleties" of the curve, $P(x)$ **must** look like this:

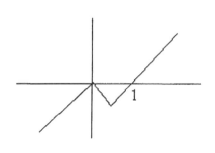

The graph, like $y = x^5$, comes up from $-\infty$ as x does. It hits 0 at $x = 0$, heads down to a minimum (CALCULUS!), comes back up to 0 at $x = 1$ and finally follows x^5 to ∞ and beyond! (Well, actually not beyond ∞, but I can't get Buzz Lightyear out of mind!) The graph is curved but I have used straight lines to show the tendencies. We find the **exact** shape of the curve (extremes, concavity) using the first and second derivatives.

Two for you.

Sketch the graphs without calculus: 1) $P(x) = (x-1)x^3(x+1)^2$ 2) $P(x) = -(x-1)^2 x^2(x+1)^2$

Answers

1)

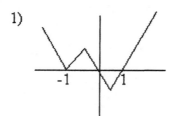

2)

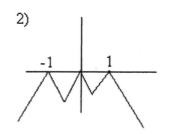

Vertical and Horizontal Asymptotes

A **vertical asymptote** is a **finite** x value $\boxed{x = \textbf{constant}}$ where $y \to \infty$ or $-\infty$. So, look for x values that make y "blow up": division by 0 or functions that have "built in" vertical asymptotes such as ln, tan, cot, sec, and csc.

A **horizontal asymptote** is a **finite** y value $\boxed{y = \textbf{constant}}$ where $x \to \infty$ or $-\infty$. Take the limit as $x \to \pm\infty$ to see if you obtain a finite y.

Example 1) Find the vertical and horizontal asymptotes for the function $y = \dfrac{1}{(x-5)(x+1)} + \ln(x-1)$

Solution $x = 5$ is a vertical asymptote. Also,

$$\lim_{x \to 1^+} \left(\frac{1}{(x-5)(x+1)} + \ln(x-1) \right) = -\frac{1}{8} + \lim_{x \to 1^+} \ln(x-1) \overset{\boxed{\lim_{x \to 1^+} \ln(x-1) = -\infty}}{=} -\infty, \text{ so } x = 1 \text{ is a vertical asymptote.}$$

Note that the function is only defined for $x > 1$, and so $x = -1$ **is not a vertical asymptote!**

$$\lim_{x \to \infty} \left(\frac{1}{(x-5)(x+1)} + \ln(x-1) \right) = 0 + \lim_{x \to \infty} \ln(x-1) = \infty, \text{ and so there are no horizontal asymptotes.}$$

Example 2) Find the vertical and horizontal asymptotes for the function $y = \dfrac{3x^2}{(x-2)(x+1)}$.

Solution The vertical asymptotes are $x = 2$ and $x = -1$. For horizontal, consider:

$$\lim_{x \to \pm\infty} \frac{3x^2}{(x-2)(x+1)} \overset{\boxed{\text{Divide top and bottom by } x^2.}}{=} \lim_{x \to \pm\infty} \frac{\left(\dfrac{3x^2}{x^2} \right)}{\left(\dfrac{(x-2)(x+1)}{x^2} \right)} = \lim_{x \to \pm\infty} \frac{3}{\left(1 - \dfrac{2}{x}\right)\left(1 + \dfrac{1}{x}\right)} = 3$$

Therefore, the horizontal asymptote is $x = 3$.

Note : **There can be more than one vertical asymptote. BUT because y is a FUNCTION, there can be only one horizontal asymptote (or none) as x approaches ∞ and one (or none) as x approaches $-\infty$.**

Note : **YOU must decide whether you need to check the limits as x approaches $+\infty$ and $-\infty$ separately!**

Two for you.

For the following, find the horizontal and vertical asymptotes.

1) $y = \dfrac{x^3}{(x-1)(x-2)}$ 2) $y = \dfrac{5x}{(3x+4)} + \dfrac{1}{e^x - 1}$

Answers 1) vertical asymptotes: $x = 1$, $x = 2$ no horizontal asymptote

2) vertical asymptotes: $x = -\dfrac{4}{3}$, $x = 0$ horizontal asymptote: $y = \dfrac{5}{3}$

Slant Asymptotes

A **slant asymptote** is a straight line that a function "asymptotically" (**what a word!**) approaches as x approaches ∞ or $-\infty$ or both. All you have to do is check the limits as $x \to \pm\infty$.

Example 1) Find the slant asymptote(s) of the function $f(x) = 2x + \dfrac{1}{x+1}$.

Solution $\lim\limits_{x \to \pm\infty} f(x)$

$$= \lim_{x \to \pm\infty}\left(2x + \frac{1}{x+1}\right) \overset{\boxed{\lim\limits_{x \to \pm\infty}\left(\frac{1}{x+1}\right)=0}}{=} \lim_{x \to \pm\infty} 2x.$$

$\therefore$ The slant asymptote is $y = 2x$.

Just for interest, $x = -1$ is a vertical asymptote.

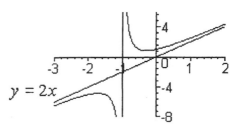

$y = 2x$

Example 2) Find the slant asymptotes, if any, of the function

$$g(x) = \frac{x + x^3}{3x^2 + 1} + e^x.$$

Solution Remember that $\lim\limits_{x \to \infty} e^x = \infty$ while $\lim\limits_{x \to -\infty} e^x = 0$.

It would therefore be prudent to check $+\infty$ and $-\infty$ separately!

$$\lim_{x \to \infty} g(x) = \lim_{x \to \infty}\left(\frac{x + x^3}{3x^2 + 1} + e^x\right) \overset{\boxed{\substack{\text{Divide top and}\\\text{bottom by } x^2.}}}{=} \lim_{x \to \infty}\left(\frac{\frac{1}{x} + x}{3 + \frac{1}{x^2}} + e^x\right) = \lim_{x \to \infty}\left(\frac{x}{3} + e^x\right)$$

$$\lim_{x \to -\infty} g(x) = \lim_{x \to -\infty}\left(\frac{x + x^3}{3x^2 + 1} + e^x\right) \overset{\boxed{\substack{\text{Divide top and}\\\text{bottom by } x^2.}}}{=} \lim_{x \to -\infty}\left(\frac{\frac{1}{x} + x}{3 + \frac{1}{x^2}} + e^x\right) \overset{\boxed{\lim\limits_{x \to -\infty} e^x = 0}}{=} \lim_{x \to -\infty}\frac{x}{3}$$

$\therefore$ There is a slant asymptote of $y = \dfrac{x}{3}$ as $x \to -\infty$ (but not $+\infty$!).

Two for you.

Find the slant asymptotes, if any, for these functions:

1) $f(x) = \dfrac{-x^5 - 2x + 7}{2x - 5x^4}$ 2) $g(x) = \dfrac{e^{-x} + 4x^2}{x - 3}$

Answers 1) $y = \dfrac{1}{5}x$ as $x \to \pm\infty$ 2) $y = 4x$ as $x \to \infty$

Intersection of Two Curves

When finding the intersection of $y = f(x)$ and $y = g(x)$, you

1) set $f(x) = g(x)$ and solve for x;
2) substitute each value of x into **either** $f(x)$ or $g(x)$ to find the corresponding y;
3) if it is fairly straightforward, sketch the graphs of the two curves so you have an idea of how many intersection points there are and approximately where to find them. **Use this as a guide but be prepared on occasion to be surprised if your guess doesn't tally with the math.**

Example 1) Find the intersection of the curves given by

$y = f(x) = x^2$ and $y = g(x) = x + 6$.

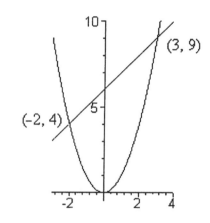

Solution Set $x^2 = x + 6$. $\therefore$ $x^2 - x - 6 = 0$. Factoring,

$(x - 3)(x + 2) = 6$ and so $x = 3$ or $x = -2$. Since $f(3) = 9$

and $f(-2) = 4$, the intersection points are $(3, 9)$ and $(-2, 4)$.

Note 1: We often forget we want the intersection points and stop once we find the x values. **Go back to ONE of the original curves to find y.**

Note 2: It doesn't matter whether you go back to f or g to find the y values. For example, here $g(3) = 9$ and $g(-2) = 4$.

Example 2) Find the intersection of $y = \sin x$ and $y = \cos x$, for $0 \le x \le 2\pi$.

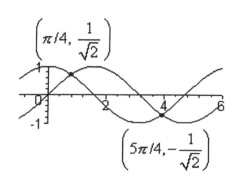

Solution Set $\sin x = \cos x$. The easiest way to tackle this one: divide both sides by $\cos x$, giving $\tan x = 1$. Now you don't need your calculator to solve this! **I hope!** In fact, if you do, your calculator would lead you astray. The calculator would give you, in radian mode, $\pi/4 \doteq 0.785$ radians. But **YOU** must realize that **tan is also positive in the third quadrant!** Don't ignore this since $0 \le x \le 2\pi$. The corresponding third quadrant angle is $5\pi/4$.

And let's not forget: $\sin\left(\dfrac{\pi}{4}\right) = \dfrac{1}{\sqrt{2}}$ and $\sin\left(\dfrac{5\pi}{4}\right) = -\dfrac{1}{\sqrt{2}}$.

The interesection points are $\left(\dfrac{\pi}{4}, \dfrac{1}{\sqrt{2}}\right)$ and $\left(\dfrac{5\pi}{4}, -\dfrac{1}{\sqrt{2}}\right)$.

Two for you.

Find the intersection of 1) $y = x^3$ and $y = x^2 + x - 1$ 2) $y = \sin x$ and $y = -\cos x,\ -\pi \le x \le \pi$

Answers 1) $(1, 1)$ and $(-1, -1)$ 2) $\left(\dfrac{3\pi}{4}, -\dfrac{1}{\sqrt{2}}\right)$ and $\left(-\dfrac{\pi}{4}, -\dfrac{1}{\sqrt{2}}\right)$

The Greatest Integer (or Floor) Function

The greatest integer function, also called the floor function, is usually denoted by $[[x]]$. **Every number** is either **an integer** or **lies between two integers, one above, one below.** The greatest integer function inputs a number.

If the number IS an integer, that's your answer! $[[4]] = 4$.

If not, your answer is the integer just below the number. $[[4.2]] = 4$.

Got it? Let's see.

Example 1) Evaluate each of the following:

(a) $[[1.7]]$ (b) $[[-2.3]]$ (c) $[[8]]$ (d) $[[-6]]$

Solution (a) 1   (b) -3 (c) 8   (d) -6

Example 2) Let $f(x) = [[x]]$, for $-2 \leq x \leq 3$. Write $f(x)$ without using greatest integer notation and draw its graph.

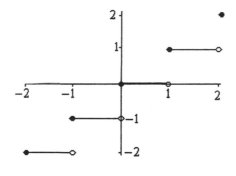

Solution $f(x) = \begin{cases} -2, & \text{if } -2 \leq x < -1 \\ -1, & \text{if } -1 \leq x < 0 \\ 0, & \text{if } 0 \leq x < 1 \\ 1, & \text{if } 1 \leq x < 2 \\ 2, & \text{if } x = 2 \end{cases}$

Example 3) Solve $[[2x+1]] = 6$.

> 2x+1 must be at least 6 but less than 7.

Solution $[[2x+1]] = 6 \iff 6 \leq 2x+1 < 7 \iff 5 \leq 2x < 6 \iff 2.5 \leq x < 3$.

Two for you.

1) Evaluate each of the following: (a) $[[7]]$ (b) $[[-2.2]]$ (c) $[[-0.0001]]$

2)(a) Solve $[[x^2]] = 1$ (Hint: solve $1 \le x^2 < 2$.)   (b) $[[x^3 - 1]] = -4$

Answers 1) (a) 7   (b) -3 (c) -1

2)(a) $-\sqrt{2} < x \le -1$ or $1 \le x < \sqrt{2}$ (b) $-3^{1/3} \le x < -2^{1/3}$

Graphs with the Greatest Integer Function

The key here is to find the equations explicitly using the appropriate intervals.

Example 1) Let $f(x) = [[2x]]$, for $-1 \le x \le 1$.
Write $f(x)$ without using greatest integer notation
and draw its graph.

Solution $f(x) = \begin{cases} -2, \text{ if } -1 \le x < -0.5 \\ -1, \text{ if } -0.5 \le x < 0 \\ 0, \text{ if } 0 \le x < 0.5 \\ 1, \text{ if } 1 \le x < 1 \\ 2, \text{ if } x = 1 \end{cases}$

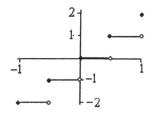

Example 2) Let $f(x) = x - [[x]]$, for $-2 \le x \le 2$.
Write $f(x)$ without using greatest integer notation
and draw its graph.

Solution $f(x) = \begin{cases} x + 2, \text{ if } -2 \le x < -1 \\ x + 1, \text{ if } -1 \le x < 0 \\ x, \text{ if } 0 \le x \\ x - 1, \text{ if } 1 \le x < 1 \\ 0, \text{ if } x = 2 \end{cases}$

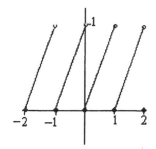

One for you.

1) Let $f(x) = 2x - [[x]]$, for $-2 \leq x \leq 1$.

Write $f(x)$ without using greatest integer notation and draw its graph.

Answer 1) $f(x) = \begin{cases} 2x+2, & \text{if } -2 \leq x < -1 \\ 2x+1, & \text{if } -1 \leq x < 0 \\ 2x, & \text{if } 0 \leq x < 1 \\ 1, & \text{if } x = 1 \end{cases}$

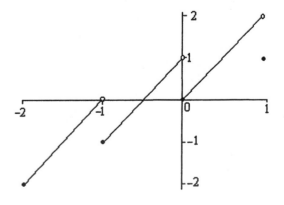

Properties of Exponents

Just as multiplication is a **short form for repeated addition** of the same number, exponentiation is a **short form for repeated multiplication** of the same number. For example, $5 \times 4 = 5 + 5 + 5 + 5$ and $5^4 = 5 \times 5 \times 5 \times 5$.

Here are the basic rules for exponents.

$$a^x a^y = a^{x+y} \qquad \frac{a^x}{a^y} = a^{x-y} \qquad (a^x)^y = a^{xy}$$

$$a^0 = 1 \qquad a^1 = a \qquad a^{-1} = \frac{1}{a}$$

$$a^{-x} = \frac{1}{a^x} \qquad \frac{1}{a^{-x}} = a^x$$

$$(ab)^x = a^x b^x \qquad \left(\frac{a^x b^y}{c^z}\right)^w = \frac{a^{xw} b^{yw}}{c^{zw}}$$

Example 1) Evaluate each of the following:

(a) 4^3 (b) 0.1^0 (c) 5^{-1} (d) 0.1^{-3} (e) $\dfrac{12}{3^{-2}}$

Solution (a) 64 (b) 1 (c) $\dfrac{1}{5}$ (d) $0.1^{-3} = \left(\dfrac{1}{10}\right)^{-3} = 10^3 = 1000$ (e) $\dfrac{12}{3^{-2}} = 12 \cdot 9 = 108$

Example 2) Simplify each of the following:

(a) $x^3 x^7$ (b) $\dfrac{z^4 z^{-2}}{z^7}$ (c) $\left(\dfrac{2^{12} 3^4}{2^{-2} 3^{11}}\right)^{-2}$

Solution (a) $x^3 x^7 = x^{10}$ (b) $\dfrac{z^4 z^{-2}}{z^7} = z^{4-2-7} = z^{-5} = \dfrac{1}{z^5}$

(c) $\left(\dfrac{2^{12} 3^4}{2^{-2} 3^{11}}\right)^{-2} = \left(\dfrac{2^{14}}{3^7}\right)^{-2} = \dfrac{2^{-28}}{3^{-14}} = \dfrac{3^{14}}{2^{28}}$

Two for you.

1) Evaluate: (a) -7^0 (b) $(-3)^3$ (c) $\dfrac{1}{3^{-4}}$

2) Simplify: (a) $x^{1/2}x^5$ (b) $\left(\dfrac{x^6 y^7}{x^{-4} y^6}\right)^5$

Answers 1)(a) -1 (b) -27 (c) 81 2)(a) $x^{11/2}$ (b) $x^{50} y^5$

Logarithms (Log Means "FIND THE EXPONENT")

Logs cause headaches. I have to admit it. Students find logs hard. Why? I think in part it's because of the word "log". It seems to have no connection with what it represents in math. Then again, students seem to find exponents easy. So when you see "log", think, even read, "**FIND THE EXPONENT!**"

For example, $\log_2 16$:

Find the exponent you need with base 2 to get a value of 16. Hmm...2 times 2 times 2 times 2...**FOUR**!

You see, that's not so hard! It's not like beating your head against a log. Sorry. Here are the rules.

$$\log_a (xy) = \log_a x + \log_a y \qquad \log_a \left(\frac{x}{y}\right) = \log_a x - \log_a y \qquad \log_a \left(x^y\right) = y \log_a x$$

$$\log_a 1 = 0 \qquad \log_a a = 1 \qquad \log_a a^{-1} = -1 \qquad \log_a \left(a^x\right) = x$$

DON'T confuse $\log_a \left(x^y\right)$ with $(\log_a x)^y$. For example,

$$\log_2 \left(4^3\right) = \log_2 \left(\left(2^2\right)^3\right) = \log_2 \left(2^6\right) = 6 \text{ while } (\log_2 4)^3 = \log_2 \left(2^2\right) \cdot \log_2 \left(2^2\right) \cdot \log_2 \left(2^2\right) = 2^3 = 8.$$

Change of base formula: $\log_a x = \dfrac{\log_b x}{\log_b a}$ and in particular, $\log_a b = \dfrac{1}{\log_b a}$

Special bases: $\log x$ **MEANS** $\log_{10} x$ $\qquad$ $\ln x$ **MEANS** $\log_e x$

Example 1) Evaluate: (a) $\log 1000$ $\qquad$ (b) $\log_2 \dfrac{1}{16}$ $\qquad$ (c) $\log_3 \left(27^4\right)$ $\qquad$ (d) $(\log_3 27)^4$

Solution (a) $\log 1000 = 3$ $\quad$ (b) $\log_2 \dfrac{1}{16} = -4$ $\quad$ (c) $\log_3 \left(27^4\right) = 4(3) = 12$ $\quad$ (d) $(\log_3 27)^4 = 3^4 = 81$

Example 2) Expand using properties of logs: (a) $\log_3 \left(\dfrac{x^3 y^4}{z^4}\right)$ $\qquad$ (b) $\log\left((x^2 + y^2)(x^2 - y^2)\right)$

Solution (a) $\log_3 \left(\dfrac{x^3 y^4}{z^5}\right) = 3\log_3 x + 4\log_3 y - 5\log_3 z$

(b) $\log\left((x^2 + y^2)(x^2 - y^2)\right) = \log(x^2 + y^2) + \log(x^2 - y^2) \overset{\boxed{\text{optional}}}{=} \log(x^2 + y^2) + \log(x - y) + \log(x + y)$

Example 3) Change $\log_9 x$ to log base 4, log base 10, and log base e.

Solution $\log_9 x = \dfrac{\log_4 x}{\log_4 9} = \dfrac{\log x}{\log 9} = \dfrac{\ln x}{\ln 9}$

93

Two for you.

1)(a) Expand using properties of logs: $\log\left(a^{-2}b\sqrt{c}\right)^3$. (Hint: rewrite $\sqrt{c}$ as $c^{1/2}$.)

  (b) Combine using properties of logs: $3\ln x - 4\ln y + 1$. (Hint: use $1 = \ln e$.)

2) Change $\log e$ (that is $\log_{10} e$) to base e.

Answers 1)(a) $-6\log a + 3\log b + \dfrac{3}{2}\log c$ (b) $\ln\left(\dfrac{ex^3}{y^4}\right)$ 2) $\log e = \dfrac{1}{\ln 10}$

Basic Exponential Graphs

Graphs with equations of the form $y = a^x$, where $a > 1$, are **really really really important**.

Example 1) Draw the graphs of

$y = 2^x$ and $y = 3^x$ on the same axes.

Solution For bases $a > 1$:

1) $\lim\limits_{x \to \infty} a^x = \infty$ and the expression

approaches ∞ **faster** as a increases.

2) $\lim\limits_{x \to -\infty} a^x = 0$ and the expression

approaches 0 **faster** as a increases.

3) $\boxed{!!!!\ a^x \ \textbf{IS ALWAYS} > 0 !!!!}$

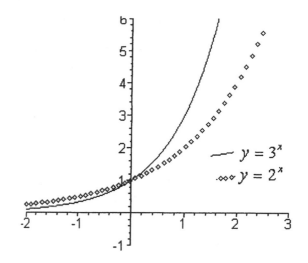

Graphs with equations of the form $y = a^x$, where $0 < a < 1$, are **really really really almost as important**.

Example 2) Draw the graphs of

$y = \left(\dfrac{1}{2}\right)^x = 2^{-x}$ and $y = \left(\dfrac{1}{3}\right)^x = 3^{-x}$ on the same axes.

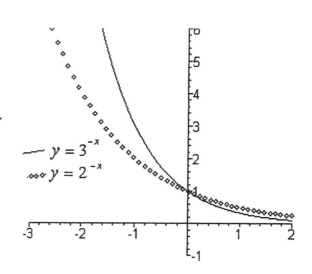

Solution

For bases $0 < a < 1$:

1) $\lim\limits_{x \to \infty} a^x = 0$ and the expression

approaches 0 **faster** as a increases.

2) $\lim\limits_{x \to -\infty} a^x = \infty$ and the expression

approaches ∞ **faster** as a increases.

3) $\boxed{!!!!\ a^x \ \textbf{IS STILL ALWAYS} > 0 !!!!}$

Two for you.

1)(a) As $x \to \infty$, which of these functions approaches ∞ faster: $y = e^x$ or $y = 10^x$?

(b) As $x \to -\infty$, which of these functions approaches 0 faster: $y = e^x$ or $y = 10^x$?

2)(a) As $x \to \infty$, which of these functions approaches 0 faster: $y = e^{-x}$ or $y = 10^{-x}$?

(b) As $x \to -\infty$, which of these functions approaches ∞ faster: $y = e^{-x}$ or $y = 10^{-x}$?

Answers 1)(a) $y = 10^x$ (b) $y = 10^x$ 2)(a) $y = 10^{-x}$ (b) $y = 10^{-x}$

Basic Logarithm Graphs

Graphs with equations of the form $y = \log_a x$, where $a > 1$, are **really really really important**.

Example 1) Draw the graphs of $y = \log_2 x$ and $y = \log_3 x$ on the same axes.

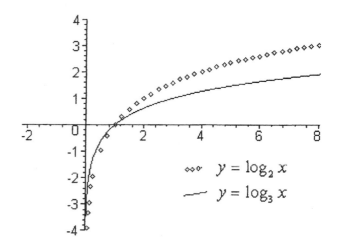

Solution For bases $a > 1$:

1) $\lim\limits_{x \to \infty} \log_a x = \infty$ and the expression

approaches ∞ **more slowly** as a increases.

2) $\lim\limits_{x \to 0^+} \log_a x = -\infty$ and the expression

approaches $-\infty$ **more slowly** as a increases.

3) !!!!To evaluate $\log_a x$, x MUST BE > 0!!!!

Graphs with equations of the form $y = \log_a x$, where $0 < a < 1$, are **really really really almost as important**.

Example 2) Draw the graphs of $y = \log_{1/2} x$ and $y = \log_{1/3} x$ on the same axes.

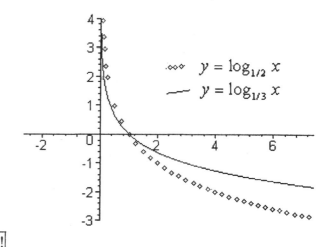

Solution For bases $0 < a < 1$:

1) $\lim\limits_{x \to 0^+} \log_a x = \infty$ and the expression

approaches ∞ **more slowly** as a **decreases**.

2) $\lim\limits_{x \to \infty} \log_a x = -\infty$ and the expression

approaches $-\infty$ **more slowly** as a **decreases**.

3) !!!!To evaluate $\log_a x$, x MUST STILL BE > 0!!!!

Two for you.

1)(a) As $x \to \infty$, which of these functions approaches ∞ faster: $y = \ln x$ or $y = \log x$?

(b) As $x \to 0^+$, which of these functions approaches $-\infty$ faster: $y = \ln x$ or $y = \log x$?

2)(a) As $x \to \infty$, which of these functions approaches $-\infty$ faster: $y = y = \log_{1/e} x$ or $y = \log_{1/10} x$?

(b) As $x \to 0^+$, which of these functions approaches ∞ faster: $y = \log_{1/e} x$ or $y = \log_{1/10} x$?

Answers 1)(a) $y = \ln x$ (b) $y = \ln x$ 2) (a) $y = \log_{1/e} x$ (b) $y = \log_{1/e} x$

Inverse Formulas for Exponents and Logarithms

Remember: Log means "FIND THE EXPONENT!" First, a review of the basics:

$$\log_a (xy) = \log_a x + \log_a y \qquad \log_a \left(\frac{x}{y}\right) = \log_a x - \log_a y \qquad \log_a (x^y) = y \log_a x$$

$$\log_a 1 = 0 \qquad \log_a a = 1 \qquad \log_a a^{-1} = -1 \qquad \log_a a^x = x$$

Change of base formula: $\log_a x = \dfrac{\log_b x}{\log_b a}$ and in particular, $\log_a b = \dfrac{1}{\log_b a}$

Special bases: $\log x$ **MEANS** $\log_{10} x$ $\ln x$ **MEANS** $\log_e x$. **And now, the...**

Inverse Formulas: $a^{\log_a x} = x \qquad 10^{\log x} = x \qquad e^{\ln x} = x$

$\log_a (a^x) = x \qquad \log(10^x) = x \qquad \ln(e^x) = x$

Example 1) Simplify the following:

(a) $\log_3 (3^{23})$ (b) $\log(10^{\sin x})$ (c) $\ln(e^{-2.37})$ (d) $\log_2 (3^x)$

Solution (a) $\log_3 (3^{23}) = 23$ (b) $\log(10^{\sin x}) = \sin x$ (c) $\ln(e^{-2.37}) = -2.37$

(d) $\log_2 (3^x) = x \log_2 3$ We can't simplify further because **the bases don't match!**

Example 2) Simplify the following:

(a) $5^{\log_5 \pi}$ (b) $10^{\log(x+5)}$ (c) $e^{\ln(\ln x)}$ (d) $7^{\log_9 x}$

Solution (a) $5^{\log_5 \pi} = \pi$ (b) $10^{\log(x+5)} = x+5$ (c) $e^{\ln(\ln x)} = \ln x$

(d) $7^{\log_9 x}$ We can't make this simpler because **the bases don't match!**

Example 3) Simplify the following:

(a) $7 \log_4 (4^{2x+5})$ (b) $5\ln(e^{x^2})$ (c) $10^{3\log 2}$ (d) $e^{7\ln x}$ (e) $\log_2 4^x$

Solution (a) $7 \log_4 (4^{2x+5}) = 7(2x+5) = 14x + 35$ (b) $5\ln(e^{x^2}) = 5x^2$ (c) $10^{3\log 2} = 10^{\log(2^3)} = 10^{\log 8} = 8$

(d) $e^{7\ln x} = e^{\ln(x^7)} = x^7$ (e) $\log_2 4^x = $ (e) $\log_2 (2^2)^x = \log_2 (2^{2x}) = 2x$

Two for you.

Simplify: 1)(a) $\log_7\left(7^{3x}\right)$ (b) $\ln(e^{4x+e^x})$ (c) $\log_2(8^t)$

2)(a) $10^{\log 7}$ (b) $e^{4\ln(\sin x)}$ (c) $4^{\log_9(x)}$

Answers 1)(a) $3x$ (b) $4x+e^x$ (c) $3t$

2)(a) 7 (b) $\sin^4(x)$ (c) No **easy** simplification since the bases are different.

Solving Exponential Equations

To solve exponential and logarithm equations, you must be completely comfortable with:

$$a = b^x \iff x = \log_b a$$

b is the **base** x is the **exponent** a is, well, we will call it the **value**.

(Actually, a is the "power" but millions and millions of people confuse "power" with "exponent"! So, let's call it the value. Humour me!)

Example 1) Find x in each of the following:

(a) $81 = 3^x$ (b) $5^{2x} = 12$ (c) $e^{x^2} = 9$ (d) $10^{4x+3} = 10^{2x-1}$ (e) $6^{3x+2} = -1$

Solution (a) $81 = 3^x \iff x = \log_3 81 = 4$

(b) $5^{2x} = 12 \iff 2x = \log_5 12 \underset{\substack{\text{optional, using the} \\ \text{change of base formula}}}{=} \frac{\ln 12}{\ln 5} \underset{\text{calculator}}{\doteq} 1.544$ and so $x \doteq 0.772$

(c) $e^{x^2} = 9 \iff x^2 = \ln 9 \iff x = \pm\sqrt{\ln 9} \overset{\text{optional}}{\doteq} \pm 1.4823$

(d) $10^{4x+3} = 10^{2x-1} \iff 4x + 3 = 2x - 1 \iff 2x = -4 \iff x = -2$

(e) $6^{3x+2} = -1$ has no solution since, as long as a is positive, then $a^x > 0$ **ALWAYS!**

Example 2) Solve for x: $e^{2x} - 2e^x - 1 = 0$

Solution Treat this as a quadratic equation with variable e^x: $\left(e^x\right)^2 - 2e^x - 1 = 0$

Now use the quadratic formula for $aX^2 + bX + c = 0$ with $a = 1$, $b = -2$, $c = -1$, and $X = e^x$

$$e^x \underset{\substack{-b\pm\sqrt{b^2-4ac} \\ \hline 2a}}{=} \frac{-(-2) \pm \sqrt{(-2)^2 - 4(1)(-1)}}{2(1)} = \frac{2 \pm \sqrt{8}}{2} \underset{\sqrt{8}=\sqrt{4\cdot2}=2\sqrt{2}}{=} \frac{2 \pm 2\sqrt{2}}{2} = 1 \pm \sqrt{2}$$

BUT $1 - \sqrt{2} < 0$ and $e^x > 0!$ $\therefore$ $e^x = 1 + \sqrt{2}$ and so $x = \ln\left(1 + \sqrt{2}\right)$

Two for you.

1) Find x in each of the following:

(a) $2^x = 128$ (b) $7^{3x+1} = 49$ (c) $e^{e^x} = e^1$ (d) $10^{\sin x} = 1$ (e) $(-6)^{3x+6} = -216$

2) Solve for x: $10^{2x} + 2(10^x) - 1 = 0$

Answers 1)(a) $x = 7$ (b) $x = \dfrac{1}{3}$ (c) $x = 0$ (d) $x = k\pi, k \in \mathbb{Z}$ (e) $x = -1$

2) $x = \log\left(-1 + \sqrt{2}\right)$ $\left(\because 10^x > 0 \therefore 10^x \neq -1 - \sqrt{2}. \right)$

Solving Logarithm Equations

To solve exponential and logarithm equations, you must be completely comfortable with:

$$a = b^x \Leftrightarrow x = \log_b a$$

b is the **base** x is the **exponent** a is, well, we will call it the **value**.

(Actually, a is the "power" but millions and millions of people confuse "power" with "exponent"! So, let's call it the value. Humour me!)

Example 1) Find x in each of the following:

(a) $\log_5(3x)=2$ (b) $\ln(x^2) = -2$ (c) $\log_3(4x-1) = \log_3(6x-3)$ (d) $\ln(-1-x^2) = 1$

Solution (a) $\log_5(3x)=2 \Leftrightarrow 3x = 5^2 = 25 \Leftrightarrow x = \dfrac{25}{3}$

(b) $\ln(x^2) = -2 \Leftrightarrow x^2 = e^{-2} = \dfrac{1}{e^2} \Leftrightarrow x = \pm\dfrac{1}{e}$

VERY TRICKY STEP!! SINCE x CAN BE <0, WE NEED TO INSERT "| |"!

OR (b) $\ln(x^2) = -2 \qquad\Leftrightarrow\qquad 2\ln|x| = -2 \Leftrightarrow \ln|x| = -1 \Leftrightarrow x = \pm\dfrac{1}{e}$

(c) $\log_3(4x-1) = \log_3(6x-3) \Leftrightarrow 4x-1 = 6x-3 \Leftrightarrow 2 = 2x \Leftrightarrow x = 1$

(d) $\ln(-1-x^2) = 1 \Leftrightarrow e = -1-x^2$. This has no solution since $e > 2$ while $-1-x^2 \le -1$.

Example 2) Solve for x: $\log x + \log(2x-1) = \log 3$

Solution Note $x > 0$ and $x > \dfrac{1}{2}$ because of the domain of log functions! So $x > \dfrac{1}{2}$ in this question.

Using log properties:

$\log x + \log(2x-1) = \log 3 \Leftrightarrow \log(x(2x-1)) = \log 3 \Leftrightarrow 2x^2 - x = 3$

$\Leftrightarrow 2x^2 - x - 3 = 0 \Leftrightarrow (2x-3)(x+1) = 0 \Leftrightarrow x = \dfrac{3}{2}$ or $x = -1$

BUT $x > \dfrac{1}{2}$, and so $x = \dfrac{3}{2}$.

Two for you.

1) Find x in each of the following:

(a) $\log_5 x = 2$ (b) $\log_2(3x) = 4$ (c) $\ln(\ln x) = 0$ (d) $\log(3x) = \log(12 - x)$

2) Solve for x: $\ln(4x + 7) - \ln(x) = \ln 5$

Answers 1)(a) $x = 25$ (b) $\dfrac{16}{3}$ (c) $x = e$ (d) $x = 3$ 2) $x = 7$

The Derivative of $y = e^x$ and $y = a^x$

I **LOVE** this formula: $\dfrac{de^x}{dx} = e^x$

> e^x **is its own derivative. The slope equals the height at each point.**

But enough of my obsession. We have examples to do and two for you!

Example 1) Find $\dfrac{dy}{dx}$ for each of the following:

(a) $y = x^3 e^x$ (b) $y = e^{\sin x}$ (c) $y = e^{4x}$ (d) $y = e^{x \ln 4}$ (e) $y = 4^x$

Solution (a) $y = x^3 e^x$ $\therefore$ $\dfrac{dy}{dx}$ [Product Rule] $= x^3(e^x) + e^x(3x^2) =$ [a little neater...] $x^2 e^x(x+3)$

(b) $y = e^{\sin x}$ $\therefore$ $\dfrac{dy}{dx}$ [Chain Rule] $= e^{\sin x} \cos x$ (c) $y = e^{4x}$ $\therefore$ $\dfrac{dy}{dx}$ [Chain Rule] $= e^{4x}(4) = 4e^{4x}$

(d) $y = e^{x \ln 4}$ $\therefore$ $\dfrac{dy}{dx}$ [Chain Rule] $= e^{x \ln 4}(\ln 4)$ $=$ [Bring the "ln 4" to the left so this answer looks like the answer in (c). Remember ln 4 is just a constant.] $(\ln 4)e^{(\ln 4)x}$

(e) $y = 4^x$. Remember a^x [$X = e^{\ln X}$. Here, $X = a^x$.] $= e^{\ln(a^x)}$ [Log Property!] $= e^{x \ln a}$. So $y = 4^x$ [Here, $a^x = 4^x$.] $= e^{\ln(4^x)} = e^{x \ln 4}$

$\therefore$ $\dfrac{dy}{dx}$ [Now this is part (d) above!] $= e^{x \ln 4}(\ln 4)$ [Don't forget: $e^{x \ln 4} = 4^x$] $= 4^x \ln 4$ [for those who like their constants at the left...] $= (\ln 4)4^x$

Part (e) gives us the rule for the derivative of $y = a^x$: $\dfrac{d(a^x)}{dx} = a^x \ln a$.

Example 2) Find $\dfrac{dy}{dx}$ for the following: (a) $y = 2^x \tan x$ (b) $y = 5^{\sec x}$ (c) $y = \dfrac{3^{x^2}}{x}$

Solution (a) $y = 2^x \tan x$ $\therefore$ $\dfrac{dy}{dx}$ [Product Rule] $= 2^x \sec^2 x + \tan x \, 2^x \ln 2$ $=$ [a little neater...] $2^x(\sec^2 x + \tan x \ln 2)$

(b) $y = 5^{\sec x}$ $\therefore$ $\dfrac{dy}{dx}$ [Chain Rule] $= 5^{\sec x} \sec x \tan x \ln 5$

(c) $y = \dfrac{3^{x^2}}{x}$ $\therefore$ $\dfrac{dy}{dx}$ [Quotient Rule] $= \dfrac{x 3^{x^2}(2x)\ln 3 - 3^{x^2}(1)}{x^2}$ [a little neater...] $= \dfrac{3^{x^2}(2x^2 \ln 3 - 1)}{x^2}$

Two for you.

Differentiate: 1)(a) $y = e^{\cos x}$ (b) $y = \sqrt{e^{2x} + 4}$

2)(a) $y = 3^{\cos x}$ (b) $y = \sqrt{a^{2x} + 4}$, where $a > 0$

Answers 1)(a) $\dfrac{dy}{dx} = -e^{\cos x} \sin x$ (b) $\dfrac{dy}{dx} = \dfrac{e^{2x}}{\sqrt{e^{2x} + 4}}$

2)(a) $\dfrac{dy}{dx} = -3^{\cos x}(\ln 3)\sin x$ (b) $\dfrac{dy}{dx} = \dfrac{a^{2x}\ln a}{\sqrt{a^{2x} + 4}}$

The Derivative of $y = \ln x$ and $y = \log_a x$

My students are pretty comfortable with the formula $\dfrac{d(\ln x)}{dx} = \dfrac{1}{x}$. But it is also true that

$\dfrac{d(\ln|x|)}{dx} = \dfrac{1}{x}$. The second formula allows $x < 0$. The good news is that the derivative is the

same! You see how the Math Gods take care of you?!

Example 1) Given $\dfrac{d(\ln x)}{dx} = \dfrac{1}{x}$, prove that $\dfrac{d(\ln|x|)}{dx} = \dfrac{1}{x}$.

Solution Case 1) Let $x > 0$ so that $|x| = x$. $\therefore$ $\dfrac{d(\ln|x|)}{dx} = \dfrac{d(\ln x)}{dx} = \dfrac{1}{x}$

Case 2) Let $x < 0$ so that $|x| = -x$. $\therefore$ $\dfrac{d(\ln|x|)}{dx} = \dfrac{d\left(\ln(-x)\right)}{dx} \overset{\text{Chain Rule!}}{=} \dfrac{-1}{(-x)} = \dfrac{1}{x}$

Example 2) Differentiate: (a) $y = \ln(3x)$ (b) $y = e^x \ln|1 + 3x|$ (c) $y = \ln(\ln x)$

Solution (a) $y = \ln(3x)$ $\therefore$ $\dfrac{dy}{dx} \overset{\boxed{\text{Chain Rule!}}}{=} \dfrac{3}{3x} = \dfrac{1}{x}$ $\quad \boxed{\text{alternate method}} \boxed{\text{log property}}$ **OR** $y = \ln 3 + \ln x$ and so $\dfrac{dy}{dx} \overset{\boxed{\text{Right away!}}}{=} \dfrac{1}{x}$

(b) $y = e^x \ln|1 + 3x|$ $\therefore$ $\dfrac{dy}{dx} \overset{\boxed{\text{Product Rule!}}}{=} e^x \dfrac{3}{1 + 3x} + \ln|1 + 3x|(e^x) = e^x\left(\dfrac{3}{1+3x} + \ln|1+3x|\right)$

(c) $y = \ln(\ln x)$ $\therefore$ $\dfrac{dy}{dx} \overset{\boxed{\text{Chain Rule!}}}{=} \dfrac{1}{\ln x}\dfrac{1}{x} \overset{\boxed{\text{a little neater...}}}{=} \dfrac{1}{x \ln x}$

Example 3) Find the derivative of $y = \log_2 x$.

Solution $y = \log_2 x \overset{\boxed{\substack{\text{Change of} \\ \text{Base formula}}}}{=} \dfrac{\log_e x}{\log_e 2} \overset{\boxed{\log_e = \ln}}{=} \dfrac{\ln x}{\ln 2} \overset{\boxed{\substack{\text{Remember ln 2} \\ \text{is a constant.}}}}{=} \dfrac{1}{\ln 2}\ln x$ $\therefore$ $\dfrac{dy}{dx} = \dfrac{1}{\ln 2}\left(\dfrac{1}{x}\right) = \dfrac{1}{x \ln 2}$

$\boxed{\text{Compare: } \dfrac{d(a^x)}{dx} = a^x \ln a \qquad \dfrac{d(\log_a x)}{dx} = \dfrac{1}{x \ln a}}$

Example 4) Differentiate: (a) $y = \log_5|1 + 3x|$ (b) $y = \log_a(\log_a x)$

Solution (a) $y = \log_5|1 + 3x|$ $\therefore$ $\dfrac{dy}{dx} \overset{\boxed{\text{Chain Rule!}}}{=} \dfrac{3}{(1 + 3x)\ln 5}$

(b) $y = \log_a(\log_a x) \overset{\boxed{\substack{\text{Be careful! Here is the} \\ \text{"outside" change of base!}}}}{=} \dfrac{\ln(\log_a x)}{\ln a}$ $\therefore$ $\dfrac{dy}{dx} \overset{\boxed{\text{Chain Rule!}}}{=} \dfrac{1}{(\log_a x)\ln a}\dfrac{1}{x \ln a} \overset{\boxed{\substack{\text{A little} \\ \text{neater...}}}}{=} \dfrac{1}{x \log_a x(\ln a)^2}$

Two for you.

Differentiate: 1) $y = \ln |\cos x| + \ln |\sec x + \tan x|$ 2) $y = \log(\log(x))$

Answers 1) $-\tan x + \sec x$ 2) $\dfrac{1}{x \log x (\ln 10)^2}$

Logarithmic Differentiation Part I

Now this is a **fun** page. Really. Here we take TOUGH looking questions and make them easy. You see, logs aren't scary. Logs are our friends. They make tough questions easy. Okay, I hear you. Let's get on with it.

Example 1) Find $\dfrac{dy}{dx}$ if $y = \ln\left(\dfrac{x^3(4x+5)^2}{(e^x+1)^5}\right)$.

Solution This looks **very scary**! Product rule, quotient rule, chain rule, power rule, log rule...**HELP**!

But... $y = \ln\left(\dfrac{x^3(4x+5)^2}{(e^x+1)^5}\right) \overset{\boxed{\text{Log Properties!}}}{=} 3\ln x + 2\ln(4x+5) - 5\ln(e^x+1)$

and so $\dfrac{dy}{dx} \overset{\boxed{\text{Don't forget the chain rule!}}}{=} \dfrac{3}{x} + \dfrac{8}{4x+5} - \dfrac{5e^x}{e^x+1}$. **EASY!**

Example 2) **Slight complication**...Find $\dfrac{dy}{dx}$ if $y = \dfrac{x^{1/3}\cos^3 x}{(x^4-x)^2}$.

Solution To use the log properties, we need to take ln of each side. But we can only take ln of **positive numbers**. So, **first, take the absolute value of each side**, that is, set |Left Side| = |Right Side
We should be concerned: does this change or restrict the original question? **NO!**

Remember, $\dfrac{d\ln|x|}{dx} \overset{\boxed{\substack{\text{Say goodbye to}\\\text{absolute value!}}}}{=} \dfrac{1}{x}$.

When we differentiate, absolute value disappears and we get the correct derivative with no restrictions!

$y = \dfrac{x^{1/3}\cos^3 x}{(x^4-x)^2}$ $\boxed{\text{Take the absolute value of each side.}}$ $\therefore$ $|y| = \left|\dfrac{x^{1/3}\cos^3 x}{(x^4-x)^2}\right| = \dfrac{|x|^{1/3}|\cos x|^3}{|x^4-x|^2}$ and so

$\ln|y| = \ln\left(\dfrac{|x|^{1/3}|\cos x|^3}{|x^4-x|^2}\right) \overset{\boxed{\text{Log Properties!}}}{=} \dfrac{1}{3}\ln|x| + 3\ln|\cos x| - 2\ln|x^4-x|$

$\boxed{\text{Note: we differentiate \textbf{implicity} on the left side.}}$
$\dfrac{1}{y}\dfrac{dy}{dx} = \dfrac{1}{3x} - \dfrac{3\sin x}{\cos x} - \dfrac{2(4x^3-1)}{x^4-x}$ and so $\dfrac{dy}{dx} \overset{\boxed{\substack{\text{Cross multiply by }y.\\\text{Also, use }\frac{3\sin x}{\cos x} = 3\tan x.}}}{=} y\left(\dfrac{1}{3x} - 3\tan x - \dfrac{2(4x^3-1)}{x^4-x}\right)$

$\boxed{\text{Optional--replace }y\text{ with the original expression.}}$
$= \dfrac{x^{1/3}\cos^3 x}{(x^4-x)^2}\left(\dfrac{1}{3x} - 3\tan x - \dfrac{2(4x^3-1)}{x^4-x}\right)$

Two for you.

Find $\dfrac{dy}{dx}$ for each of the following: 1) $y = \ln\left(\dfrac{x^5 \ln x}{\tan x}\right)$ 2) $y = \left(\dfrac{(5x-1)\sin^5 x}{e^x + 5}\right)^3$

Answers 1) $\dfrac{5}{x} + \dfrac{1}{x \ln x} - \dfrac{\sec^2 x}{\tan x}$ 2) $3y\left(\dfrac{5}{5x-1} + 5\cot x - \dfrac{e^x}{e^x + 5}\right)$

Logarithmic Differentiation Part II: $\frac{d}{dx}\left(f(x)^{g(x)}\right)$

You know how to find $\frac{dy}{dx}$ if $y = (f(x))^3$: $\frac{dy}{dx} = 3(f(x))^2 f'(x)$. This is "variable^{constant}".

You know how to find $\frac{dy}{dx}$ if $y = 3^{f(x)}$: $\frac{dy}{dx} = 3^{f(x)} f'(x) \ln 3$. This is "constant^{variable}".

But what about $\frac{dy}{dx}$ if $y = f(x)^{g(x)}$, that is, "variable^{variable}"?

Example 1) Differentiate $y = (2x+1)^{\sin x}$.

Solution Here we use the same method as Log Differentiation Part I. But because exponential functions are defined only when the base is positive...

> We allow the **arithmetic** expression $(-3)^3$ but we don't allow the **function** $y = (-3)^x$!

...we don't need to take the absolute value of each side first.

$$y = (2x+1)^{\sin x} \quad \therefore \quad \ln y = \ln\left((2x+1)^{\sin x}\right) = \sin x \ln(2x+1).$$

$$\therefore \quad \frac{1}{y}\frac{dy}{dx} \overset{\boxed{\text{Product Rule}}}{=} \sin x \left(\frac{2}{2x+1}\right) + \ln(2x+1)(\cos x) \quad \text{and so} \quad \frac{dy}{dx} = y\left(\left(\frac{2\sin x}{2x+1}\right) + \ln(2x+1)(\cos x)\right)$$

$$\overset{\boxed{\text{Optional: replace } y.}}{=} (2x+1)^{\sin x}\left(\left(\frac{2\sin x}{2x+1}\right) + \ln(2x+1)(\cos x)\right) \overset{\boxed{\text{Get a common denominator.}}}{=} (2x+1)^{\sin x}\left(\frac{2\sin x + (2x+1)\ln(2x+1)(\cos x)}{2x+1}\right)$$

$$\overset{\boxed{\begin{array}{l}\text{Notice the exponential base } 2x+1:\\ \text{exponent is } \sin x \text{ on top and 1 on the bottom!}\end{array}}}{=} (2x+1)^{\sin x - 1}\left(2\sin x + (2x+1)\ln(2x+1)(\cos x)\right)$$

Example 2) Find $\frac{dy}{dx}$ if $y = (x^2+1)^{\cos x}$.

Solution $\ln y = \ln\left((x^2+1)^{\cos x}\right) = \cos x \ln(x^2+1)$.

$$\therefore \quad \frac{1}{y}\frac{dy}{dx} = \cos x\left(\frac{2x}{x^2+1}\right) + \ln(x^2+1)(-\sin x) \quad \text{and so} \quad \frac{dy}{dx} = y\left(\left(\frac{2x\cos x}{x^2+1}\right) - \sin x \ln(x^2+1)\right)$$

$$= (x^2+1)^{\cos x}\left(\left(\frac{2x\cos x}{x^2+1}\right) - \sin x \ln(x^2+1)\right) \overset{\boxed{\text{optional}}}{=} (x^2+1)^{\cos x - 1}\left(2x\cos x - (x^2+1)\sin x \ln(x^2+1)\right)$$

Two for you.

Differentiate: 1) $y = x^{5x+1}$ 2) $y = (\tan x)^{\ln x}$

Answers 1) $\dfrac{dy}{dx} = x^{5x+1}\left(\dfrac{5x+1}{x} + 5\ln x\right) \overset{\boxed{\text{optional}}}{=} x^{5x}(5x+1+5x\ln x)$

2) $\dfrac{dy}{dx} = (\tan x)^{\ln x}\left(\dfrac{\ln x \sec^2 x}{\tan x} + \dfrac{\ln(\tan x)}{x}\right) \overset{\boxed{\text{optional}}}{=} (\tan x)^{\ln x-1}\left(\dfrac{\left(x\ln x \sec^2 x\right) + \tan x \ln(\tan x)}{x}\right)$

112

Integrals Yielding ln: $\int \dfrac{\left(\frac{du}{dx}\right)}{u}dx = \ln|u|+C$

Compare these three integrals:

1) $\int \dfrac{1}{x^2+1}dx$ 2) $\int \dfrac{2x}{(x^2+1)^2}dx$ 3) $\int \dfrac{2x}{x^2+1}dx$. Which one fits into the pattern $\int \dfrac{\left(\frac{du}{dx}\right)}{u}dx$?

In all three examples, $u = x^2+1$.

Integral 1) misses the pattern because $\dfrac{du}{dx} = 2x$ is nowhere to be found.

Integral 2) misses the pattern because the exponent on u in the bottom is 2.

Integral 3) is like Goldilock's choice of porridge: just right!

In fact, $\int \dfrac{1}{x^2+1}dx = \arctan x + C$, which you may not have yet studied!

In fact, $\int \dfrac{2x}{(x^2+1)^2}dx$ is really the Chain Rule in Reverse with the Power Rule $\int u^{-2}\dfrac{du}{dx}dx$:

$$\int \dfrac{2x}{(x^2+1)^2}dx = \int (x^2+1)^{-2}\,2x\,dx = \dfrac{(x^2+1)^{-1}}{-1}+C = \dfrac{-1}{x^2+1}+C$$

But $\int \dfrac{2x}{x^2+1}dx = \ln|x^2+1|+C$ $\boxed{x^2+1>0 \text{ so we don't need absolute value.}}$ $= \ln(x^2+1)+C$

Example 1) Evaluate the following integrals:

(a) $\int \dfrac{1+\sin x}{x-\cos x}dx$ (b) $\int \dfrac{e^{3x}}{e^{3x}-5}dx$ (c) $\int \dfrac{1}{x\ln x}dx$

Solution (a) $\int \dfrac{1+\sin x}{x-\cos x}dx = \ln|x-\cos x|+C$

(b) $\int \dfrac{e^{3x}}{e^{3x}-5}dx$ $\boxed{\text{Adjust the multiplicative constant.}}$ $= \dfrac{1}{3}\int \dfrac{3e^{3x}}{e^{3x}-5}dx = \dfrac{1}{3}\ln|e^{3x}-5|+C$

(c) $\int \dfrac{1}{x\ln x}dx$ $\boxed{\text{Here, } u=\ln x \text{ and } \frac{du}{dx}=\frac{1}{x}.}$ $= \int \dfrac{\left(\frac{1}{x}\right)}{\ln x}dx = \ln|\ln x|+C$

Two for you.

Evaluate the following integrals: 1) $\int \dfrac{x^2}{2x^3+1}\,dx$ 2) $\int \dfrac{1}{x \ln x \ln(\ln x)}\,dx$ $\left(\text{Hint: Let } u = \ln(\ln x).\right)$

Answers 1) $\dfrac{1}{6}\ln|2x^3+1| + C$ 2) $\ln|\ln(\ln x)| + C$

Radian Measure of an Angle

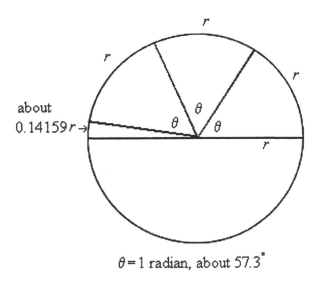

about
$0.14159r \rightarrow$

$\theta = 1$ radian, about $57.3°$

Take any circle. Take the circle's radius r. Wrap r around the circumference. This will create an angle just a little over 57 degrees. We call this 1 radian. If you wrap another radius around the circle starting at the end of the first wrapped r, and then another, and then another, you will find that the semi-circle will have 3 **and about** 0.14159 radii wrapped around it in all. We give the number of radii wrapped around the semi-circle the name π. So

$\pi \doteq 3.14159$ and π radians $= 180°$.

It is customary (Don't ask me why!) to omit the word "radians" but not "degrees"!

Example 1) Convert the following radian measures to degrees:

(a) $\dfrac{\pi}{2}$ (b) $\dfrac{7\pi}{3}$ (c) 2 r r r r r

Solution Since π radians $= 180°$, $\therefore$ 1 radian $= \left(\dfrac{180}{\pi}\right)°$. From now on, we will omit "radians".

(a) $\dfrac{\pi}{2} = \dfrac{\pi}{2}\left(\dfrac{180}{\pi}\right)° = 90°$ (b) $\dfrac{7\pi}{3} = \dfrac{7\pi}{3}\left(\dfrac{180}{\pi}\right)° = 420°$

(c) $2 = 2\left(\dfrac{180}{\pi}\right)° = \left(\dfrac{360}{\pi}\right)° \doteq 2(57.3)° = 114.6°$

Example 2) Convert the following degree measures to radians:

(a) $45°$ (b) $-40°$ (c) $\pi°$

Solution Since $180° = \pi$ radians, $\therefore$ 1 degree $= \left(\dfrac{\pi}{180}\right)$ radians. Again, we will now omit "radians'

(a) $45° = 45\left(\dfrac{\pi}{180}\right) = \dfrac{\pi}{4}$ (b) $-40° = -40\left(\dfrac{\pi}{180}\right) = -\dfrac{2\pi}{9}$ (c) $\pi° = \pi\left(\dfrac{\pi}{180}\right) = \dfrac{\pi^2}{180} \doteq 0.055$

Two for you.

1) Convert to degree measure: (a) $\pi/6$ (b) -1.8

2) Convert to radian measure: (a) $-60°$ (b) $12°$

Answers 1)(a) $30°$ (b) $\left(\dfrac{-324}{\pi}\right)° \doteq -103.1$ 2)(a) $-\dfrac{\pi}{3}$ (b) $\dfrac{12\pi}{180} = \dfrac{\pi}{15} \doteq 0.21$

Angles in Standard Position

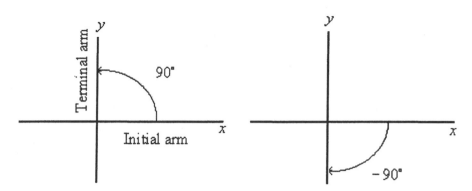

POSITIVE angles are drawn **COUNTER-CLOCKWISE** from the positive x axis.

NEGATIVE angles are drawn **CLOCKWISE** from the positive x axis

Example 1)(a) Give, in degree measure, **all** the angles which have the same initial and terminal arms as (i) $90°$ (ii) $-90°$, and   (b) draw both $450°$ and $-450°$ in standard position.

Solution (a)(i) Let k be a **positive** integer. Then all the angles $90°, 450°, 810°$, and in general, $(90 + 360k)°$ have the same initial and terminal arms. Also, $90°, -270°, -630°$, and in general, $(90 - 360k)°$ have the same initial and terminal arms.

(ii) $-90°, -450°, -810°$, and in general, $(-90 - 360k)°$ all have the same initial and terminal arms. Also, $-90°, 270°, 630°$, and in general, $(-90 + 360k)°$ have the same initial and terminal arms.

SUMMARY : (i) $(90 + 360k)°$, for **any integer** k, all have the same initial and terminal arms as $90°$.
(ii) $(-90 + 360k)°$, for **any integer** k, all have the same initial and terminal arms as $-90°$.

(b)

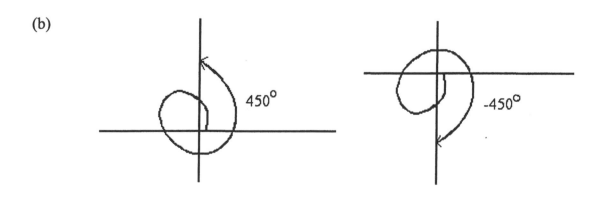

117

Two for you.

1) Give all angles in standard position which have the same terminal arm as (a) $45°$ (b) -60

2) Redo question 1, giving your answers in radian measure.

Answers 1)(a) $(45 + 360k)°$, for $k \in \mathbb{Z}$ (b) $(-60 + 360k)°$, for $k \in \mathbb{Z}$

2)(a) $\dfrac{\pi}{4} + 2k\pi$, for $k \in \mathbb{Z}$ (b) $-\dfrac{\pi}{3} + 2k\pi$, for $k \in \mathbb{Z}$

Similar Triangles

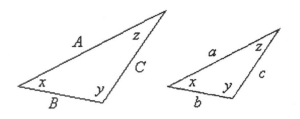

Two triangles are similar if all three pairs of corresponding angles are equal and corresponding sides are in the same ratio.   In this pair of similar triangles $\nearrow$

$\dfrac{A}{a} = \dfrac{B}{b} = \dfrac{C}{c}$. Note that side "$A$" in the big triangle and side "a" in the small triangle

correspond because they are both between angles x and z.

Conditions for similarity: We can conclude two triangles are similar if we know

SSS : all three pairs of corresponding sides are in the same ratio **or**

SAS : two pairs of corresponding sides are in the same ratio and the contained angles are equal **or**

AAA : all three pairs of corresponding angles are equal **or**

AA : if two pairs of angles are equal, the third pair must be as well, so the triangles are similar by AAA

Example 1) In the diagram at the top, if $A = 10$, $B = 8$, and $a = 4$, find C, b, and c.

Solution $\dfrac{A}{a} = \dfrac{10}{4} = \dfrac{5}{2}$. Therefore, $\dfrac{B}{b} = \dfrac{8}{b} = \dfrac{5}{2}$ and so $b = \dfrac{16}{5}$. As for c and C, we know that

$\dfrac{C}{c} = \dfrac{5}{2}$ and so $C = \dfrac{5c}{2}$. However, without knowing either C or c (or at least one of the angles),

we can not find either the value of C or c! (Don't you hate trick questions?!)

Example 2) In the picture to the right, find a formula for s in terms of d.

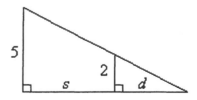

Solution By similar triangles (AAA), $\dfrac{s+d}{d} = \dfrac{5}{2}$. Therefore, $2s + 2d = 5d$ and so $s = \dfrac{3d}{2}$.

Two for you.

1) In the triangles at the right, suppose $A = 7$, $B = 4$, $b = 2$, and $c = 3$. Find a, and C.

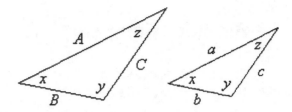

2) Use the similar triangles below to find an expression for a in terms of b.

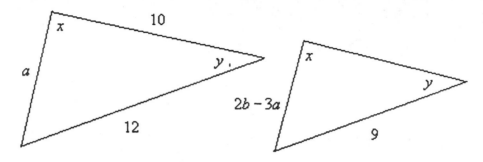

Answers 1) $a = 7/2$ and $C = 6$ 2) $a = 8b/15$

120

Trig Ratios for the $(30°, 60°, 90°) \equiv (\pi/6, \pi/3, \pi/2)$ Triangle

You DON'T have to memorize these ratios. It is soooooooooooooooooooooooooooo easy to take 30 seconds and redevelop them!

Draw an **equilateral** (and therefore **equiangular**) triangle. Each angle is $60° = \pi/3$ radians. Drop a perpendicular from one vertex. This divides the triangle into two congruent triangles, with angles of $30°$, $60°$, and $90°$ or, in radians, $\pi/6$, $\pi/3$, and $\pi/2$. Let each side in the original triangle have length 2. Then in each of the two congruent triangles, the sides **(using congruency and Pythagoras!)** are 2, 1, and $\sqrt{3}$.

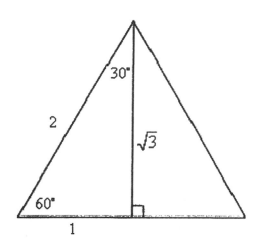

For $30°$, 1 is **opposite** and $\sqrt{3}$ is **adjacent**.
For $60°$, $\sqrt{3}$ is **opposite** and 1 is **adjacent**.
For both, 2 is the **hypotenuse.**

SOHCAHTOA!
SIN=O/H
COS=A/H
TAN=O/A
CSC=H/O
SEC=H/A
COT=A/O

(Important! Do this construction yourself at least THREE TIMES!!)

Example 1) State the sine, cosine and tangent ratios for $30°$ and $60°$.

Solution $\quad \sin(30°) = \sin\left(\dfrac{\pi}{6}\right) = \dfrac{1}{2} \qquad \cos(30°) = \cos\left(\dfrac{\pi}{6}\right) = \dfrac{\sqrt{3}}{2} \qquad \tan(30°) = \tan\left(\dfrac{\pi}{6}\right) = \dfrac{1}{\sqrt{3}}$

$\qquad\qquad\quad \sin(60°) = \sin\left(\dfrac{\pi}{3}\right) = \dfrac{\sqrt{3}}{2} \qquad \cos(60°) = \cos\left(\dfrac{\pi}{3}\right) = \dfrac{1}{2} \qquad \tan(60°) = \tan\left(\dfrac{\pi}{3}\right) = \sqrt{3}$

121

Two for you.

1) State the cosecant, secant, and cotangent ratios for 30° and 60°.

2) Is it a coincidence that $\sin(30°) = \cos(60°)$?

Answers 1) $\csc(30°) = \csc\left(\dfrac{\pi}{6}\right) = 2$ $\sec(30°) = \sec\left(\dfrac{\pi}{6}\right) = \dfrac{2}{\sqrt{3}}$ $\cot(30°) = \cot\left(\dfrac{\pi}{6}\right) = \sqrt{3}$

$\csc(60°) = \csc\left(\dfrac{\pi}{3}\right) = \dfrac{2}{\sqrt{3}}$ $\sec(60°) = \sec\left(\dfrac{\pi}{3}\right) = 2$ $\cot(60°) = \cot\left(\dfrac{\pi}{3}\right) = \dfrac{1}{\sqrt{3}}$

2) **NO! Opposite** for 30° is **adjacent** for 60°! In fact, if a and b are two **complementary** angles in a triangle (that is, they add to 90°), then $\sin a = \cos b$, $\cos a = \sin b$, and $\tan a = \cot b$.

Trig Ratios for the $(45°, 45°, 90°) \equiv (\pi/4, \pi/4, \pi/2)$ Triangle

You DON'T have to memorize these ratios. It is soooooooooooooooooooooooooooo easy to take 30 seconds and redevelop them!

Draw a right isosceles triangle, that is, a draw a 90° angle and make the two attached sides equal.

Therefore, the triangle has angles of 45°, 45°, and 90° or in radians, $\pi/4$, $\pi/4$, and $\pi/2$.

Let the two equal sides have length 1. Then (**using Pythagoras!**), the hypotenuse has length $\sqrt{2}$.

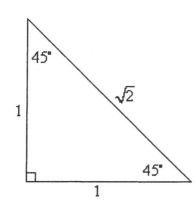

For 45°, 1 is **opposite** and 1 is **adjacent**.
The **hypotenuse** is $\sqrt{2}$.

SOHCAHTOA!
SIN=O/H
COS=A/H
TAN=O/A
CSC=H/O
SEC=H/A
COT=A/O

Example 1) State the sine, cosine and tangent ratios for 45°.

Solution $\quad \sin(45°) = \sin\left(\dfrac{\pi}{4}\right) = \dfrac{1}{\sqrt{2}} \qquad \cos(45°) = \cos\left(\dfrac{\pi}{4}\right) = \dfrac{1}{\sqrt{2}} \qquad \tan(45°) = \tan\left(\dfrac{\pi}{4}\right) = 1$

Example 2) Given that $\sin^2(\theta) = \dfrac{1 - \cos(2\theta)}{2}$, find the exact value of $\sin\left(\dfrac{\pi}{8}\right)$.

Solution Taking $\theta = \dfrac{\pi}{8}$, $\quad \sin^2\left(\dfrac{\pi}{8}\right) = \dfrac{1 - \cos\left(\dfrac{\pi}{4}\right)}{2} = \dfrac{1 - \dfrac{1}{\sqrt{2}}}{2} \overset{\boxed{\text{Get a common denominator ON THE TOP and simplify.}}}{=} \dfrac{\sqrt{2}-1}{2\sqrt{2}} \overset{\boxed{\text{Some of us love to rationalize the bottom!}}}{=} \dfrac{2-\sqrt{2}}{4}$

and so $\sin\left(\dfrac{\pi}{8}\right) \overset{\boxed{\text{Take the POSITIVE root since the angle is in the first quadrant.}}}{=} \sqrt{\dfrac{2-\sqrt{2}}{4}} \doteq 0.383$

123

Two for you.

1) State the cosecant, secant, and cotangent ratios for 45°.

2) Using $\cos^2(\theta) = \dfrac{1 + \cos(2\theta)}{2}$, find the exact value of $\cos\left(\dfrac{\pi}{8}\right)$.

Answers 1) $\csc(45°) = \sec(45°) = \sqrt{2}$ $\cot(45°) = 1$ 2) $\cos\left(\dfrac{\pi}{8}\right) = \sqrt{\dfrac{2 + \sqrt{2}}{4}}$

Trig Ratios for $30°, 45°, 60°$ (and More) – A Table!

degrees	radians	sin	cos	tan	csc	sec	cot
0	0	0	1	0	undefined	1	undefined
30	$\dfrac{\pi}{6}$	$\dfrac{1}{2}$	$\dfrac{\sqrt{3}}{2}$	$\dfrac{1}{\sqrt{3}}$	2	$\dfrac{2}{\sqrt{3}}$	$\sqrt{3}$
45	$\dfrac{\pi}{4}$	$\dfrac{1}{\sqrt{2}}$	$\dfrac{1}{\sqrt{2}}$	1	$\sqrt{2}$	$\sqrt{2}$	1
60	$\dfrac{\pi}{3}$	$\dfrac{\sqrt{3}}{2}$	$\dfrac{1}{2}$	$\sqrt{3}$	$\dfrac{2}{\sqrt{3}}$	2	$\dfrac{1}{\sqrt{3}}$
90	$\dfrac{\pi}{2}$	1	0	undefined	1	undefined	0
120	$\dfrac{2\pi}{3}$	$\dfrac{\sqrt{3}}{2}$	$-\dfrac{1}{2}$	$-\sqrt{3}$	$\dfrac{2}{\sqrt{3}}$	-2	$-\dfrac{1}{\sqrt{3}}$
135	$\dfrac{3\pi}{4}$	$\dfrac{1}{\sqrt{2}}$	$-\dfrac{1}{\sqrt{2}}$	-1	$\sqrt{2}$	$-\sqrt{2}$	-1
150	$\dfrac{5\pi}{6}$	$\dfrac{1}{2}$	$-\dfrac{\sqrt{3}}{2}$	$-\dfrac{1}{\sqrt{3}}$	2	$-\dfrac{2}{\sqrt{3}}$	$-\sqrt{3}$
180	π	0	-1	0	undefined	-1	undefined
210	$\dfrac{7\pi}{6}$	$-\dfrac{1}{2}$	$-\dfrac{\sqrt{3}}{2}$	$\dfrac{1}{\sqrt{3}}$	-2	$-\dfrac{2}{\sqrt{3}}$	$\sqrt{3}$
225	$\dfrac{5\pi}{4}$	$-\dfrac{1}{\sqrt{2}}$	$-\dfrac{1}{\sqrt{2}}$	1	$-\sqrt{2}$	$-\sqrt{2}$	1
240	$\dfrac{4\pi}{3}$	$-\dfrac{\sqrt{3}}{2}$	$-\dfrac{1}{2}$	$\sqrt{3}$	$-\dfrac{2}{\sqrt{3}}$	-2	$\dfrac{1}{\sqrt{3}}$
270	$\dfrac{3\pi}{2}$	-1	0	undefined	-1	undefined	0
300	$\dfrac{5\pi}{3}$	$-\dfrac{\sqrt{3}}{2}$	$\dfrac{1}{2}$	$-\sqrt{3}$	$-\dfrac{2}{\sqrt{3}}$	2	$-\dfrac{1}{\sqrt{3}}$
315	$\dfrac{7\pi}{4}$	$-\dfrac{1}{\sqrt{2}}$	$\dfrac{1}{\sqrt{2}}$	-1	$-\sqrt{2}$	$\sqrt{2}$	-1
330	$\dfrac{11\pi}{6}$	$-\dfrac{1}{2}$	$\dfrac{\sqrt{3}}{2}$	$-\dfrac{1}{\sqrt{3}}$	-2	$\dfrac{2}{\sqrt{3}}$	$-\sqrt{3}$
360	2π	0	1	0	undefined	1	undefined

Two for you.

In **absolute value**, the numbers (or ratios) in the 1) 30° row   2) 45° row
are the same as ...

Answers 1) 150°, 210°, 330° rows   2) 135°, 225°, 315° rows

Trig Ratios for $30°, 45°, 60°$ (and More) – A (Fabulous) Picture!!

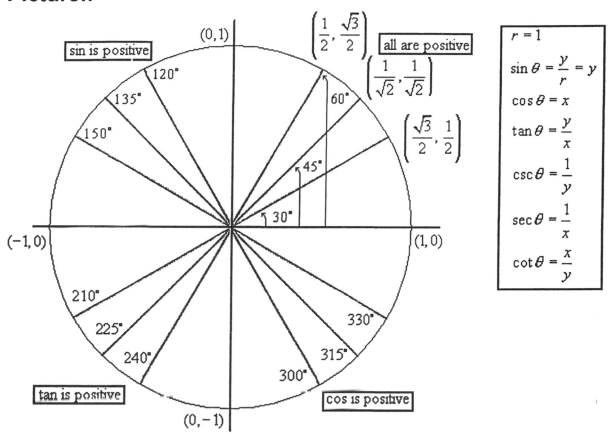

Example 1) In the above **fabulous** picture, look at the triangle formed by drawing a perpendicular from the 60° point to the x axis. Name the second, third, and fourth quadrant angles (between 90° and 360°) that give congruent triangles by drawing perpendiculars to the x axis.

Solution second quadrant: 120°; third quadrant: 240°; fourth quadrant: 300°

Example 2) Find, using the above **fabulous** picture, (a) sin(300°)   (b) tan(180°).

Solution (a) 300° is a **fourth quadrant** angle, so the sine is **negative**. The corresponding first quadrant point is $\left(\dfrac{1}{2}, \dfrac{\sqrt{3}}{2}\right)$. Therefore, $\sin(300°) = -\dfrac{\sqrt{3}}{2}$.

(b) Using the point $(-1, 0)$, $\tan(180°) = \dfrac{y}{x} = \dfrac{0}{-1} = 0$.

Two for you.

Find: 1) $\cos(135°)$ 2) $\cot(-180°)$

Answers 1) $-\dfrac{1}{\sqrt{2}}$ 2) undefined

Basic Trigonometric Graphs

$y = \sin x$

Domain $= \mathbb{R}$

Range $= [-1, 1]$

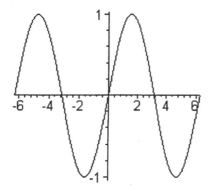

$y = \cos x$

Domain $= \mathbb{R}$

Range $= [-1, 1]$

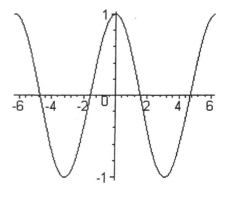

$y = \tan x$

Domain: $x \in \mathbb{R}$,

$x \neq \dfrac{\pi}{2} + k\pi, \ k \in \mathbb{Z}$

Range $= \mathbb{R}$

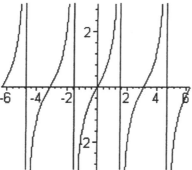

$y = \cot x$

Domain: $x \in \mathbb{R}$,

$x \neq k\pi, \ k \in \mathbb{Z}$

Range $= \mathbb{R}$

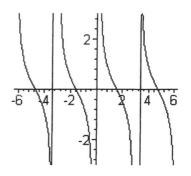

$y = \csc x$

Domain: $x \in \mathbb{R}$,

$x \neq k\pi, \ k \in \mathbb{Z}$

Range $=$

$(-\infty, -1] \cup [1, \infty)$

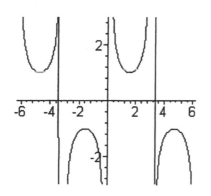

$y = \sec x$

Domain: $x \in \mathbb{R}$,

$x \neq \dfrac{\pi}{2} + k\pi, \ k \in \mathbb{Z}$

Range $=$

$(-\infty, -1] \cup [1, \infty)$

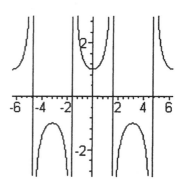

Two for you.

1) Using "...", write the restrictions on the domain for the tangent and secant functions.

2) Using "...", write the restrictions on the domain for the cotangent and cosecant functions.

Answers 1) $\quad ...-\dfrac{3\pi}{2}, -\dfrac{\pi}{2}, \dfrac{\pi}{2}, \dfrac{3\pi}{2},...$ $\qquad$ 2) $\quad ...-2\pi, -\pi, 0, \pi, 2\pi, ...$

The Circle Definition of Sine and Cosine

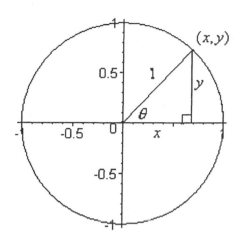

This is the circle $x^2 + y^2 = 1$. In the picture, θ is between 0 and $\pi/2$.

$$\sin\theta = \frac{y}{1} = y \quad \text{and} \quad \cos\theta = \frac{x}{1} = x$$

Mathematicians, being sensible people, said, "Let θ be any angle. Draw this angle in standard postion (counter-clockwise from the positive x axis for positive angles, clockwise for negative). The terminal arm will puncture (**OUCH!**) the circle at a point (x, y)."

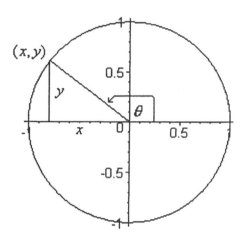

The mathematicians then said, "Let $\sin\theta = y$ and $\cos\theta = x$. The result: before, we had only sine and cosine for angles between 0 and $\pi/2$. Now, we have trig ratios for **all angles.** For any angle, the sine is the y value and the cosine is the x. So, as θ goes from 0 to $\pi/2$ to π to $3\pi/2$ to 2π, the y value, that is, $\sin\theta$, goes from 0 to 1 to 0 to -1 to 0. The x value, that is, $\cos\theta$, goes from 1 to 0 to -1 to 0 to 1. From 2π to 4π, these patterns repeat. From 0 to -2π, they repeat in reverse. And that is all there really is to the sine and cosine functions."

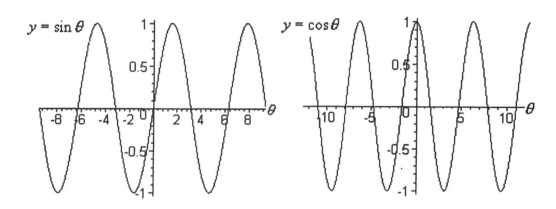

131

One for you.

1) Use the circle definition of the trigonometric functions to explain the **CAST RULE**.

$$\begin{array}{c|c}
S & A \\
\hline
T & C
\end{array}$$

Answer 1) For example, in the second quadrant $(\pi/2 < \theta < \pi)$ where $x < 0$ and $y > 0$, $\cos\theta = x$ is negative, $\sin\theta = y$ is positive, and $\tan\theta = \dfrac{y}{x}$ is negative. This explains the "S" in CA$\boxed{\text{S}}$T.

Solving the Trig Equation $\sin x = c$

DangerDangerDangerDangerDangerDangerDangerDanger

If you ask your calculator (or math processor such as Maple or Mathematica) for help solving the equation $\sin x = c$ for x, **the calculator thinks this is the sine function!** ↗

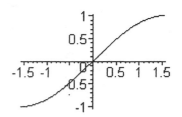

For each choice of c between -1 and 1, the calculator thinks, "Okay, there is exactly **ONE** value of x between $-\pi/2$ and $\pi/2$ that satisfies $\sin x = c$ and that value of x is the answer!" So your calculator gives you an answer in the **first quadrant when** $c \geq 0$, and the **fourth quadrant when** $c \leq 0$.

But you and I know that $\sin x = c$ **has an infinite number of solutions.** Let's call the calculator answer the **principal** solution. Ask, when you solve $\sin x = c$, "Do I want the principal solution, another solution (for example, an answer in a different quadrant), or **all** possible solutions?"

Example 1) Solve, **using your knowledge of basic trig ratios (not a calculator).** Give, in radians, the principal solution, **a (not "the"!)** solution in the other **appropriate** quadrant, and the general solution.
(a) $\sin x = 0.5$ (b) $\sin x = -0.5$

Solution (a) $\sin x = 0.5$ $\therefore$ $x = \dfrac{\pi}{6}$ Using the CAST rule, sine is also positive in the second quadrant.

A **(not "the")** corresponding second quadrant angle is $\pi - \pi/6 = 5\pi/6$.

The general solution: $x = \dfrac{\pi}{6} + 2k\pi$ or $x = \dfrac{5\pi}{6} + 2k\pi$, for $k \in \mathbb{Z}$

(b) $\sin x = -0.5$ $\therefore$ $x = -\dfrac{\pi}{6}$ Using the CAST rule, sine is also negative in the third quadrant.

A **(not "the"!)** corresponding third quadrant angle is $\pi + \pi/6 = 7\pi/6$.

The general solution: $x = -\dfrac{\pi}{6} + 2k\pi$ or $x = \dfrac{7\pi}{6} + 2k\pi$, for $k \in \mathbb{Z}$

Example 2) Find the principal solution to (a) $\sin x = 0.1$ (b) $\sin x = -0.3$.

Solution (a) $\boxed{\textbf{MOST CALCULATORS: 0.1 second function sin} =}$ $x \doteq 5.74°$ or 0.1 radians

(b) $\boxed{\textbf{MOST CALCULATORS: 0.3} \pm \textbf{ second function sin} =}$ $x \doteq -17.5°$ or -0.305 radians

Two for you.

1) Find, in degree measure, the principal, the "other quadrant", and the general solution to $\sin x = 1/$

2) Find the principal and third quadrant solutions in radians to $\sin x = -0.9$.

(Hint: for the third quadrant, use $\pi + |$ principal solution $|$.)

Answers

1) principal solution: $x = 45°$; second quadrant: $x = 135°$;

general: $x = 45° + 360k$ or $x = 135° + 360k, k \in \mathbb{Z}$

2) principal solution: $x \doteq -1.1$; third quadrant: $x \doteq 4.3$

Solving the Trig Equation $\cos x = c$

DangerDangerDanger DangerDangerDangerDangerDanger

If you ask your calculator (or math processor such as Maple or Mathematica) for help in solving the equation $\cos x = c$ for x, **the calculator thinks this is the cosine function!** ↗

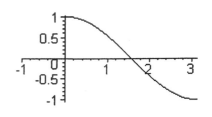

For each choice of c between -1 and 1, the calculator thinks, "Okay, there is exactly **ONE** value of x between 0 and π that satisfies $\cos x = c$ and **that** value of x is the answer!" So your calculator gives you an answer in the **first quadrant when $c \geq 0$,** and the **second quadrant when $c < 0$**

But you and I know that $\cos x = c$ has an infinite number of solutions. Let's call the calculator answer the **principal** solution. Ask, when you solve $\cos x = c$, "Do I want the principal solution, another solution (for example, an answer in a different quadrant), or **all** possible solutions?"

Example 1) Solve, **using your knowledge of basic trig ratios (not a calculator).** Give the principal solution, **a (not "the"!)** solution in the **other appropriate quadrant**, and the general solution.

(a) $\cos x = 0.5$ (b) $\cos x = -0.5$

Solution (a) $\cos x = 0.5$ $\therefore$ $x = \dfrac{\pi}{3}$ Using the CAST rule, cosine is also positive in the fourth quadrant. A (**not "the"!**) corresponding fourth quadrant angle is $-\pi/3$.

The general solution: $x = \dfrac{\pi}{3} + 2k\pi$ or $x = -\dfrac{\pi}{3} + 2k\pi$, for $k \in \mathbb{Z}$

(b) $\cos x = -0.5$ $\therefore$ $x = \dfrac{2\pi}{3}$ Using the CAST rule, cosine is also negative in the third quadrant. A (**not "the"!**) corresponding third quadrant angle is $2\pi - 2\pi/3 = 4\pi/3$.

The general solution: $x = \dfrac{2\pi}{3} + 2k\pi$ or $x = \dfrac{4\pi}{3} + 2k\pi$, for $k \in \mathbb{Z}$

Example 2) Find the principal solution to (a) $\cos x = 0.1$ (b) $\cos x = -0.3$.

Solution (a) $\boxed{\textbf{MOST CALCULATORS: 0.1 second function cos} =}$ $x \doteq 84.3°$ or 1.47 radians

(b) $\boxed{\textbf{MOST CALCULATORS: 0.3} \pm \textbf{ second function cos} =}$ $x \doteq 107.5°$ or 1.88 radians

Two for you.

1) Find, in degree measure, the principal, the "other quadrant", and the general solution to $\cos x = 1/$

2) Find the principal and third quadrant solutions in radians to $\cos x = -0.9$.

(Hint: for the third quadrant, use $2\pi - |\text{principal solution}|$.)

Answers

1) principal solution: $x = 45°$;   fourth quadrant: $x = -45°$;

general: $x = 45° + 360k$ or $x = -45° + 360k, \ k \in \mathbb{Z}$

2) principal solution: $x \doteq 2.69$; third quadrant: $x \doteq 3.59$

The Sine Law

Check whether your calculator is in degree or radian measure!

According to the Sine Law, in the triangle at the right,

$$\frac{a}{\sin A} = \frac{b}{\sin B} = \frac{c}{\sin C}.$$

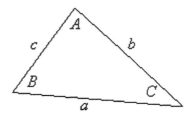

To use the Sine Law, you need two sides and the **non-contained angle** or any two angles and one side. (Remember: if you have two angles, you have three!) In the case of two sides and the contained angle, you can use the Cosine Law. So life is good!

Example 1) If in the above triangle, $a = 12$, $b = 10$, $A = \dfrac{\pi}{3}$, and $C = \dfrac{\pi}{7}$, find B (radians) and c

Solution $\dfrac{\sin B}{b} = \dfrac{\sin A}{a} \boxed{A=\frac{\pi}{3}\text{ and }a=12} = \dfrac{\left(\dfrac{\sqrt{3}}{2}\right)}{12}$ and so $\sin B \boxed{b=10} = \dfrac{10\sqrt{3}}{24} \doteq 0.722$

$\boxed{\text{MOST CALCULATORS: 0.722 second funtion sin}}$

Therefore, $B \qquad = \qquad 0.81$ radians.

Also, $\dfrac{c}{\sin C} = \dfrac{a}{\sin A} = \dfrac{12}{\left(\dfrac{\sqrt{3}}{2}\right)}$ and so $c = \dfrac{24\sin\left(\pi/7\right)}{\sqrt{3}} \doteq 6.01$.

Example 2) There are two possible triangles where $b = 11$, $c = 6$, and $C = 30°$, as illustrated at right. Find, in degrees, the **acute** and **obtuse** values of B.

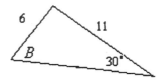
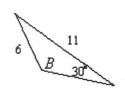

Solution $\dfrac{\sin B}{11} = \dfrac{\sin 30°}{6}$ and so $\sin B = \dfrac{(1/2)(11)}{6} = \dfrac{11}{12}$.

Therefore, $B \overset{\boxed{\text{If } B \text{ is acute!}}}{\doteq} 66.4°$ **OR** $B \overset{\boxed{\text{If } B \text{ is obtuse!}}}{\doteq} 180° - 66.4° = 113.6°$.

One for you.

1) In the triangle at the right, $a = 3$, $c = 5$, and $A = 30°$.

Find in degrees the two possible values of C.

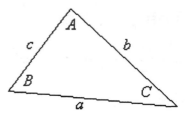

Answer 1) C $\stackrel{.}{=}$ 56.44° **OR** C $\stackrel{.}{=}$ 123.56°

$\boxed{\text{If } C \text{ is acute!}}$ $\boxed{\text{If } C \text{ is obtuse!}}$

The Cosine Law

| **Check whether your calculator is in degree or radian measure!** |

According to the Cosine Law, in the triangle at the right,

$a^2 = b^2 + c^2 - 2bc\cos A,$

with similar formulas for b and c. Also,

$\cos A = \dfrac{b^2 + c^2 - a^2}{2bc}$, again with similar

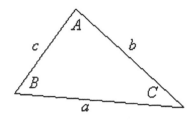

formulas for $\cos B$ and $\cos C$.

So, using the Cosine Law, to find a side you need the other two sides and the contained angle. (Compare the Sine Law). To find an angle, you need all three sides.

Example 1) In the above triangle, $a = 10$, $b = 12$, and $C = \dfrac{\pi}{7}$. Find c.

Solution $c^2 = 10^2 + 12^2 - 2(10)(12)\cos\left(\dfrac{\pi}{7}\right) \doteq 244 - 240(0.901) = 27.77$

and so $c \doteq 5.27$.

Example 2) Find, in degree measure, A in this triangle.

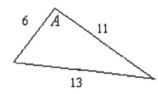

Solution $\cos A = \dfrac{11^2 + 6^2 - 13^2}{2 \cdot 11 \cdot 6} \doteq -0.0909$.

| MOST CALCULATORS: 0.0909 ± second function cos |

Therefore, $A \doteq 95.2°$.

Note: with the Cosine Law, if the required angle is obtuse (that is, between $90°$ and $180°$), the calculator gives us the correct value. There is no possible ambiguity as can happen with the Sine Law. (See The Sine Law, Example 2.)

One for you.

1) In this triangle, use the Cosine Law to find side b and then, in degrees, A.

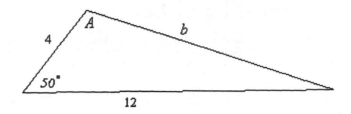

Answer 1) $b \doteq 9.9$ and $A \doteq 112°$

Commonly Used Trigonometric Identities Including Derivatives and Integrals

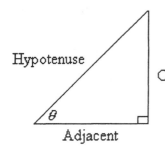

$$\sin\theta = \frac{\text{Opposite}}{\text{Hypotenuse}} = \frac{1}{\csc\theta} = \text{SOH}$$

$$\cos\theta = \frac{\text{Adjacent}}{\text{Hypotenuse}} = \frac{1}{\sec\theta} = \text{CAH}$$

$$\tan\theta = \frac{\text{Opposite}}{\text{Adjacent}} = \frac{1}{\cot\theta} = \text{TOA}$$

$$\boxed{\text{SOHCAHTOA}}$$

$\sin(-\theta) = -\sin\theta \qquad \cos(-\theta) = \cos\theta \qquad \tan(-\theta) = -\tan\theta$

$\cos^2\theta + \sin^2\theta = 1 \qquad 1 + \tan^2\theta = \sec^2\theta \qquad \cot^2\theta + 1 = \csc^2\theta$

$\sin(A \pm B) = \sin A \cos B \pm \cos A \sin B \qquad \cos(A \pm B) = \cos A \cos B \mp \sin A \sin B$

$\sin(2A) = 2\sin A \cos B \qquad \cos(2A) = \cos^2 A - \sin^2 A$

$$\tan(A \pm B) = \frac{\tan A \pm \tan B}{1 \mp \tan A \tan B} \qquad \tan(2A) = \frac{2\tan A}{1 - \tan^2 A}$$

$$\sin^2\theta = \frac{1 - \cos(2\theta)}{2} \qquad \cos^2\theta = \frac{1 + \cos(2\theta)}{2}$$

$$\frac{d(\sin\theta)}{d\theta} = \cos\theta \qquad \int \sin\theta\, d\theta = -\cos\theta + C \quad \text{(where } C \text{ is a constant)}$$

$$\frac{d(\cos\theta)}{d\theta} = -\sin\theta \qquad \int \cos\theta\, d\theta = \sin\theta + C$$

$$\frac{d(\tan\theta)}{d\theta} = \sec^2\theta \qquad \int \tan\theta\, d\theta = -\ln|\cos\theta| + C$$

$$\frac{d(\csc\theta)}{d\theta} = -\csc\theta\,\cot\theta \qquad \int \csc\theta\, d\theta = \ln|\csc\theta - \cot\theta| + C$$

$$\frac{d(\sec\theta)}{d\theta} = \sec\theta\,\tan\theta \qquad \int \sec\theta\, d\theta = \ln|\sec\theta + \tan\theta| + C$$

$$\frac{d(\cot\theta)}{d\theta} = -\csc^2\theta \qquad \int \cot\theta\, d\theta = \ln|\sin\theta| + C$$

Two for you.

How can you derive the formulas 1) $1 + \tan^2 \theta = \sec^2 \theta$ and 2) $\cot^2 \theta + 1 = \csc^2 \theta$ from $\cos^2 \theta + \sin^2 \theta = 1$?

Answers 1) Divide the formula $\cos^2 \theta + \sin^2 \theta = 1$ by $\cos^2 \theta$.

2) Divide the formula $\cos^2 \theta + \sin^2 \theta = 1$ by $\sin^2 \theta$.

Easy Limits: "No Problem" Problems

When evaluating $\lim_{x \to a} f(x)$, **YOU MUST NOT LET** x **EQUAL** a. You consider instead the "behaviour" of the function $f(x)$ as x gets closer and closer to a. However, we all know that as long as the function is "well-behaved", we do, in practice, just substitute a in for x in $f(x)$

Example 1) Evaluate the following limits:

(a) $\lim_{x \to 3} x^2$ (b) $\lim_{x \to \frac{\pi}{4}} \sin x$ (c) $\lim_{x \to 3} \dfrac{x^2 - 1}{\ln x}$

Solution (a) $\lim_{x \to 3} x^2 = 9$ (b) $\lim_{x \to \frac{\pi}{4}} \sin x = \sin\left(\dfrac{\pi}{4}\right) = \dfrac{1}{\sqrt{2}}$ (c) $\lim_{x \to 3} \dfrac{x^2 - 1}{\ln x} = \dfrac{8}{\ln 3}$

Example 2) Let $g(x) = \begin{cases} x^2 - 1, & \text{if } x < -4 \\ x^3, & \text{if } x \geq -4 \end{cases}$.

Find (a) $\lim_{x \to -6} g(x)$ (b) $\lim_{x \to -2} g(x)$ (c) $\lim_{x \to -4.01} g(x)$ (d) $\lim_{x \to -3.99} g(x)$

Solution (a) $\lim_{x \to -6} g(x) \overset{\boxed{x < -4}}{=} 35$ (b) $\lim_{x \to -2} g(x) \overset{\boxed{x > -4}}{=} -8$

(c) $\lim_{x \to -4.01} g(x) \overset{\boxed{x \text{ is still LESS than} -4.}}{=} (-4.01)^2 - 1 = 15.081$

(d) $\lim_{x \to -3.99} g(x) \overset{\boxed{x \text{ is still GREATER than} -4.}}{=} (-3.99)^3 = 63.521199$

We have to be careful with $g(x)$ only if we consider the limit as x approaches **EXACTLY** -4.

Example 3) Evaluate: (a) $\lim_{x \to 2.3} [[x]]$ (b) $\lim_{x \to -3.1} [[x]]$

Solution (a) $\lim_{x \to 2.3} [[x]] = [[2.3]] = 2$ (b) $\lim_{x \to -3.1} [[x]] = [[-3.1]] = -4$

We must be careful with the greatest integer function $[[x]]$ only when x is approaching an integer.

Two for you.

1) Evaluate the following limits:

(a) $\lim\limits_{x\to\pi} \cos(2x)$ (b) $\lim\limits_{x\to-4.2} [[x]]$ (c) $\lim\limits_{x\to3} \dfrac{|x-5|}{x-5}$

2) Let $f(x) = \begin{cases} e^x, \text{if } x \neq 2 \\ 5, \text{ if } x = 2 \end{cases}$. Find $\lim\limits_{x\to2} f(x)$.

Answers 1)(a) 1   (b) -5 (c) -1 2) e^2

"0/0" Limits

We can't divide by 0. But depending on how the top and bottom approach 0 compared to one another, we can often evaluate limits that yield "0/0" using direct (**read naïve!!**) substitution.

Example 1) Evaluate the following limits:

(a) $\lim\limits_{x \to 3}\left(\dfrac{x^2 - 9}{x - 3} \right)$ 　　(b) $\lim\limits_{x \to -1}\left(\dfrac{x^2 + 2x + 1}{x^3 + 1} \right)$ 　　(c) $\lim\limits_{x \to 16}\left(\dfrac{\sqrt{x} - 4}{x - 16} \right)$ 　　(d) $\lim\limits_{x \to -125}\left(\dfrac{x + 125}{x^{1/3} + 5} \right)$

Solution (a) $\lim\limits_{x \to 3}\left(\dfrac{x^2 - 9}{x - 3} \right)$ $\overset{\boxed{\text{Factor and divide out the common factor.}\\ \text{Say } \textbf{GOODBYE} \text{ to the problem "0/0"!}}}{=}$ $\lim\limits_{x \to 3}\left(\dfrac{(\cancel{x - 3})(x + 3)}{\cancel{x - 3}} \right) = \lim\limits_{x \to 3}(x + 3) = 6$

(b) $\lim\limits_{x \to -1}\left(\dfrac{x^2 + 2x + 1}{x^3 + 1} \right) = \lim\limits_{x \to -1}\left(\dfrac{(x + 1)^2}{(x + 1)(x^2 - x + 1)} \right) = \lim\limits_{x \to -1}\left(\dfrac{x + 1}{x^2 - x + 1} \right) = \dfrac{0}{3} = 0$

(c) $\lim\limits_{x \to 16}\left(\dfrac{\sqrt{x} - 4}{x - 16} \right)$ $\overset{\boxed{\text{Rationalize the top using}\\ \text{difference of squares:}\\ (a-b)(a+b)=a^2-b^2,\\ \text{where } a=\sqrt{x} \text{ and } b=4.}}{=}$ $\lim\limits_{x \to 16}\left(\dfrac{\sqrt{x} - 4}{x - 16} \right)\left(\dfrac{\sqrt{x} + 4}{\sqrt{x} + 4} \right) = \lim\limits_{x \to 16}\left(\dfrac{x - 16}{(x - 16)(\sqrt{x} + 4)} \right) = \lim\limits_{x \to 16}\left(\dfrac{1}{\sqrt{x} + 4} \right) = \dfrac{1}{8}$

OR

(c) $\lim\limits_{x \to 16}\left(\dfrac{\sqrt{x} - 4}{x - 16} \right)$ $\overset{\boxed{\text{Factor the bottom using difference of squares.}}}{=}$ (c) $\lim\limits_{x \to 16}\left(\dfrac{\sqrt{x} - 4}{(\sqrt{x} - 4)(\sqrt{x} + 4)} \right) = \lim\limits_{x \to 16}\left(\dfrac{1}{\sqrt{x} + 4} \right) = \dfrac{1}{8}$

(d) $\lim\limits_{x \to -125}\left(\dfrac{x + 125}{x^{1/3} + 5} \right)$ $\overset{\boxed{\text{Factor the top using sum of cubes:}\\ a^3+b^3 \text{ where } a=x^{1/3} \text{ and } b=5.}}{=}$ $\lim\limits_{x \to -125}\left(\dfrac{\left(x^{1/3} + 5\right)\left(x^{2/3} - 5x^{1/3} + 25\right)}{x^{1/3} + 5} \right)$

$= \lim\limits_{x \to -125}\left(x^{2/3} - 5x^{1/3} + 25 \right) = 75$

Three for you.

Evaluate the following limits:

1) $\lim\limits_{x \to 10}\left(\dfrac{x^2 - 100}{x - 10}\right)$ 2) $\lim\limits_{x \to -2}\left(\dfrac{x^3 + 8}{x^2 - 4}\right)$ 3) $\lim\limits_{x \to -27}\left(\dfrac{x^{1/3} + 3}{x + 27}\right)$

Answers 1)(a) 20   2) -3 3) $\dfrac{1}{27}$

One-sided Limits

If a function changes its definition at a point, we must check the limits from the left and the right separately.

Example 1) Let $f(x) = \begin{cases} 3, & \text{if } x < 1 \\ 0, & \text{if } x = 1 \\ x+2, & \text{if } x > 1 \end{cases}$ and $g(x) = \begin{cases} \cos(\pi x), & \text{if } x < 1 \\ 5, & \text{if } x = 1 \\ \sin(\pi x), & \text{if } x > 1 \end{cases}$. Evaluate the following:

(a) $\lim\limits_{x \to 1^-} f(x)$　　(b) $\lim\limits_{x \to 1^+} f(x)$　　(c) $\lim\limits_{x \to 1} f(x)$　　d) $\lim\limits_{x \to 1^-} g(x)$　　(e) $\lim\limits_{x \to 1^+} g(x)$　　(f) $\lim\limits_{x \to 1} g(x)$

Solution (a) $\lim\limits_{x \to 1^-} f(x) \overset{\boxed{x<1}}{=} \lim\limits_{x \to 1^-}(3) = 3$

(b) $\lim\limits_{x \to 1^+} f(x) \overset{\boxed{x>1}}{=} \lim\limits_{x \to 1^+}(x+2) = 3$

(c) $\lim\limits_{x \to 1} f(x) = 3$ since the left-hand limit and the right-hand limit are both equal to 3.

(d) $\lim\limits_{x \to 1^-} g(x) \overset{\boxed{x<1}}{=} \lim\limits_{x \to 1^-} \cos(\pi x) = \cos(\pi) = -1$　　(e) $\lim\limits_{x \to 1^+} g(x) \overset{\boxed{x>1}}{=} \lim\limits_{x \to 1^+} \sin(\pi x) = \sin(\pi) = 0$

(f) $\lim\limits_{x \to 1} g(x)$ does not exist since the left-hand limit $\neq$ the right-hand limit.

Note that $f(1) = 0$ and $g(1) = 5$ and **neither of these values play ANY part in the limits!**

Sometimes, YOU must realize that one-sided limits are necessary!

Example 2) Evaluate the following limits:

(a) $\lim\limits_{x \to 2} \dfrac{|x-2|}{x-2}$　　(b) $\lim\limits_{x \to 4}(x + [[x]])$

Solution (a) $\lim\limits_{x \to 2^+} \dfrac{|x-2|}{x-2} \overset{\boxed{x>2 \text{ so } x-2>0 \\ \therefore |x-2| = x-2}}{=} \lim\limits_{x \to 2^+} \dfrac{x-2}{x-2} = \lim\limits_{x \to -2^+} 1 = 1$

$\lim\limits_{x \to 2^-} \dfrac{|x-2|}{x-2} \overset{\boxed{x<2 \text{ so } x-2<0 \\ \therefore |x-2| = -(x-2)}}{=} \lim\limits_{x \to 2^-} \dfrac{-(x-2)}{x-2} = \lim\limits_{x \to 2^-}(-1) = -1$　$\therefore$　$\lim\limits_{x \to 2} \dfrac{|x-2|}{x-2}$ does not exist.

(b) $\lim\limits_{x \to 4^+}(x + [[x]]) \overset{\boxed{4<x<5 \text{ so} \\ [[x]] = 4}}{=} \lim\limits_{x \to 4^+}(x+4) = 8$　　$\lim\limits_{x \to 4^-}(x + [[x]]) \overset{\boxed{3<x<4 \text{ so} \\ [[x]] = 3}}{=} \lim\limits_{x \to 4^-}(x+3) = 7$

$\therefore$　$\lim\limits_{x \to 4}(x + [[x]])$ does not exist.

Two for you.

1) Let $f(x) = \begin{cases} x^2, & \text{if } x < -4 \\ x^3, & \text{if } x \ge -4 \end{cases}$. Find $\lim\limits_{x \to -4^-} f(x)$.

2) Evaluate: (a) $\lim\limits_{x \to -3^+} \dfrac{x+3}{|x+3|}$ (b) $\lim\limits_{x \to -5^-} (x \cdot [[x]])$

Answers 1) 16   2)(a) 1   (b) 30

Limits which Approach $\pm\infty$

Whenever a limit yields the form "$\dfrac{\text{constant}}{0}$" where the constant **IS NOT 0**, the answer **WILL BE INFINITE**. We just need to determine if it is $+\infty$ or $-\infty$.

Keep these basic infinite limits in mind:
$$\lim_{x\to 0^+}\frac{1}{x}=+\infty \quad \text{and} \quad \lim_{x\to 0^-}\frac{1}{x}=-\infty$$

Also, remember that some functions have infinite limits "built in". For example, $\lim\limits_{x\to 0^+}\dfrac{1}{\ln x}=-\infty$.

Example 1) Evaluate (a) $\lim\limits_{x\to 3^+}\left(\dfrac{x^2+1}{x-3}\right)$ (b) $\lim\limits_{x\to 3^-}\left(\dfrac{x^2+1}{x-3}\right)$

Solution (a) Note that just substituting yields "$\dfrac{10}{0}$". Since x is **greater than and approaching 3**, therefore $x-3$ is **POSITIVE** and **SMALL**. $\therefore$ $\lim\limits_{x\to 3^+}\left(\dfrac{x^2+1}{x-3}\right)$ [See the personal note below to explain where the "10" comes from!] $=$ $10\lim\limits_{x\to 3^+}\left(\dfrac{1}{x-3}\right)=+\infty$

A personal note : in these questions, I (yes me, the author!) like to evaluate everything except the factor that is creating the "0 denominator" and pull those results OUTSIDE the limit.

(b) $\lim\limits_{x\to 3^-}\left(\dfrac{x^2+1}{x-3}\right)=10\lim\limits_{x\to 3^-}\left(\dfrac{1}{x-3}\right)=-\infty$

Example 2) Evaluate $\lim\limits_{x\to 4^-}\left(\dfrac{1-x^2}{(4-x)(x+1)}\right)$.

Solution $\lim\limits_{x\to 4^-}\left(\dfrac{1-x^2}{(4-x)(x+1)}\right)$ [Evaluate everything but the $4-x$ term and rewrite it as $-(x-4)$.] $=$ $\dfrac{-15}{5}\lim\limits_{x\to 4^-}\left(\dfrac{1}{-(x-4)}\right)$ [Why? It's easier to look at $x-4$ than $4-x$ as $x\to 4^-$.] $=$ $3\lim\limits_{x\to 4^-}\left(\dfrac{1}{x-4}\right)$ [$\because x<4$ $\therefore x-4<0$] $=-\infty$

Two for you.

Find the following limits: 1) (a) $\lim\limits_{x \to 1^+} \dfrac{1}{\ln(x-1)}$ (b) $\lim\limits_{x \to \frac{\pi}{2}^-} \tan x$

2) $\lim\limits_{x \to 5^-} \left(\dfrac{x^2 - 25}{(x-5)^2} \right)$ (Hint: factor and simplify first.)

Answers 1)(a) $-\infty$ (b) $+\infty$ 2) $-\infty$

Limits At Infinity

Keep these basic limits "at infinity" (that is, limits where x approaches $+\infty$ or $-\infty$) in mind:

$$\lim_{x\to+\infty}\frac{1}{x}=0 \quad \text{and} \quad \lim_{x\to-\infty}\frac{1}{x}=0$$

Some functions have limits at infinity "built in". For example, $\lim_{x\to-\infty}2^x=0$.

Often, these kinds of questions arise when you have $\dfrac{\text{(almost) a polynomial}}{\text{(almost) another polynomial}}$.

In these examples, the **EASIEST** method is to divide the top and the bottom by the **HIGHEST POWER OF** x **IN THE** <u>**BOTTOM**</u>!

Example 1) Evaluate: (a) $\lim_{x\to\infty}\left(\dfrac{x+1}{3x+3}\right)$ (b) $\lim_{x\to-\infty}\left(\dfrac{x^2+\sin x}{x^3+2x}\right)$ (c) $\lim_{x\to-\infty}\left(\dfrac{x^3-x+\cos x}{x^2+1+\sin x}\right)$

Solution (a) $\lim_{x\to\infty}\left(\dfrac{x+1}{3x+2}\right)\overset{\boxed{\text{Divide top and bottom by }x.}}{=}\lim_{x\to\infty}\dfrac{\left(\dfrac{x+1}{x}\right)}{\left(\dfrac{3x+2}{x}\right)}=\lim_{x\to\infty}\left(\dfrac{1+\dfrac{1}{x}}{3+\dfrac{2}{x}}\right)=\dfrac{1+0}{3+0}=\dfrac{1}{3}$

(b) $\lim_{x\to-\infty}\left(\dfrac{x^2+\sin x}{x^3+2x}\right)\overset{\boxed{\text{Divide top and bottom by }x^3.}}{=}\lim_{x\to-\infty}\left(\dfrac{\dfrac{1}{x}+\dfrac{\sin x}{x^3}}{1+\dfrac{2}{x^2}}\right)\overset{\boxed{\begin{array}{l}\text{Remember }-1\le\sin x\le 1\text{ so }\frac{\sin x}{x^3}\\ \text{is, in magnitude, }\frac{\text{small number}}{\text{BIG NUMBER}}\\ \text{and will therefore approach 0!}\end{array}}}{=}\dfrac{0+0}{1+0}=\dfrac{0}{1}=0$

(c) $\lim_{x\to-\infty}\left(\dfrac{x^3-x+\cos x}{x^2+1+\sin x}\right)\overset{\boxed{\begin{array}{l}\text{Divide top}\\ \text{and bottom by}\\ x^2\text{, NOT }x^3!\end{array}}}{=}\lim_{x\to-\infty}\left(\dfrac{x-\dfrac{1}{x}+\dfrac{\cos x}{x^2}}{1+\dfrac{1}{x^2}+\dfrac{\sin x}{x^2}}\right)\overset{\boxed{\begin{array}{l}\text{Watch the special way we handle}\\ \text{the }x\text{ on top! Evaluate each limit}\\ \text{except the one that still goes to }\pm\infty.\end{array}}}{=}\dfrac{\lim_{x\to-\infty}x-0+0}{1+0+0}=-\infty$

Two for you.

Evaluate the limits:

1) $\lim\limits_{x \to -\infty} \left(\dfrac{4x^3 - 2x + e^x}{3x^3 - 5x^2 + \sin x} \right)$

2) $\lim\limits_{x \to \infty} \left(\dfrac{x+1}{\sqrt{4x^2 + x} + 5x} \right)$

$\left(\text{Hint: the highest power of } x \text{ in the bottom is } x, \text{ because of the } \text{``}\sqrt{}\text{''.} \right.$

$\left. \text{Another hint: } \dfrac{\sqrt{4x^2 + x}}{x} = \dfrac{\sqrt{4x^2 + x}}{\sqrt{x^2}} = \sqrt{\dfrac{\text{you get}}{\text{the idea!}}} \right)$

Answers 1) $\dfrac{4}{3}$ 2) $\dfrac{1}{7}$

An "$\infty - \infty$" Limit: $\lim\limits_{x \to \infty}\left(\sqrt{x^2 + 8x} - x\right)$

You might think "$\infty - \infty$" has to be 0. Could be, but not necessarily.

Example 1) Evaluate $\lim\limits_{x \to \infty}\left(\sqrt{x^2 + 8x} - x\right)$.

Solution $\lim\limits_{x \to \infty}\left(\sqrt{x^2 + 8x} - x\right) \overset{\boxed{\begin{array}{l}\text{Rationalize the numerator}\\ \text{using } (a-b)(a+b)=a^2-b^2,\\ \text{with } a=\sqrt{x^2+8x} \text{ and } b=x.\end{array}}}{=} \lim\limits_{x \to \infty}\left(\sqrt{x^2 + 8x} - x\right)\left(\frac{\sqrt{x^2 + 8x} + x}{\sqrt{x^2 + 8x} + x}\right)$

$= \lim\limits_{x \to \infty}\left(\frac{x^2 + 8x - x^2}{\sqrt{x^2 + 8x} + x}\right) = \lim\limits_{x \to \infty}\left(\frac{8x}{\sqrt{x^2 + 8x} + x}\right) \overset{\boxed{\begin{array}{l}\text{Divide the top and the bottom by } x.\\ \text{Use } x=\sqrt{x^2} \text{ (which is true when } x>0)\\ \text{to divide } x \text{ into } \sqrt{x^2+8x}.\end{array}}}{=} \lim\limits_{x \to \infty}\left(\frac{8}{\sqrt{1 + \dfrac{8}{x}} + 1}\right) = \frac{8}{2} = 4$

BUT...

Example 2) Evaluate $\lim\limits_{x \to \infty}\left(\sqrt{x^2 + 8} - x\right)$

Solution $\lim\limits_{x \to \infty}\left(\sqrt{x^2 + 8} - x\right) \overset{\boxed{\begin{array}{l}\text{Rationalize the numerator}\\ \text{using } (a-b)(a+b)=a^2-b^2.\end{array}}}{=} \lim\limits_{x \to \infty}\left(\sqrt{x^2 + 8} - x\right)\left(\frac{\sqrt{x^2 + 8} + x}{\sqrt{x^2 + 8} + x}\right)$

$= \lim\limits_{x \to \infty}\left(\frac{x^2 + 8 - x^2}{\sqrt{x^2 + 8} + x}\right) = \lim\limits_{x \to \infty}\left(\frac{8}{\sqrt{x^2 + 8} + x}\right) \overset{\boxed{\begin{array}{l}\text{Divide the top and the}\\ \text{bottom by } x. \text{ Use } x=\sqrt{x^2}.\end{array}}}{=} \lim\limits_{x \to \infty}\left(\frac{\dfrac{8}{x}}{\sqrt{1 + \dfrac{8}{x}} + 1}\right) = \frac{0}{2} = 0$

153

Two for you.

1) Evaluate $\lim_{x \to \infty}\left(\sqrt{x^2 - x + 1} - x\right)$.

2) Evaluate $\lim_{x \to -\infty}\left(\sqrt{x^2 - 3x} + x\right)$.

$$\left(\begin{array}{l} \text{Hint: for } x < 0, \ \sqrt{x^2} = -x. \text{ For example, if } x = -10, \text{ then } \sqrt{(-10)^2} = \sqrt{100} = 10 = -(-10). \\[2mm] \text{Another hint: } \frac{\sqrt{x^2 - 3x}}{x} \overset{\boxed{\substack{\text{For } x<0,\\ x=-\sqrt{x^2}.}}}{=} \frac{\sqrt{x^2 - 3x}}{-\sqrt{x^2}} = -\sqrt{\frac{x^2 - 3x}{x^2}} = -\sqrt{1 - \frac{3}{x}} \end{array}\right)$$

Answers 1) $-\dfrac{1}{2}$ 2) $\dfrac{3}{2}$

Limits: A Summary

This is a non-standard Survival Kit page but my students find this summary of limit questions REALLY useful. Almost every limit question you encounter will look like one of these!

Case 1) "No Problem" Problems. Just letting x get close to a leads to a clear answer.

Example 1) $\lim\limits_{x \to 6}\left(\dfrac{x^2 - 25}{x + 5}\right) = \dfrac{11}{11} = 1$

Case 2) "$\dfrac{c}{0}$", where $c \neq 0$. Here, the answer will be $+\infty$ or $-\infty$, or possibly both,

in which case you need to consider one-sided limits.

Example 2) $\lim\limits_{x \to -5^+}\left(\dfrac{x^2}{x + 5}\right) = 25 \lim\limits_{x \to -5^+}\left(\dfrac{1}{x + 5}\right)^{\boxed{x+5>0}} = +\infty$

Case 3) "$\dfrac{0}{0}$" when $x = a$. Divide out $\dfrac{x - a}{x - a}$, possibly more than once, and the

problem should reduce to case 1 or 2.

Example 3) $\lim\limits_{x \to -5}\left(\dfrac{x^2 - 25}{x + 5}\right) = \lim\limits_{x \to -5}\left(\dfrac{(x - 5)\,\cancel{(x + 5)}}{\cancel{x + 5}}\right) = -10$

Case 4) One-sided limits. Any of cases 1, 2, and 3 can come in the form of one-sided limits. These can be either explicit or require you to consider separate cases.

Example 4) Explicit: $\lim\limits_{x \to 2^-}[[x]] = 1;$

Implicit: $\lim\limits_{x \to 2}[[x]]$ $\because$ $\lim\limits_{x \to 2^-}[[x]] = 1 \neq \lim\limits_{x \to 2^+}[[x]] = 2,$ $\therefore$ $\lim\limits_{x \to 2}[[x]]$ does not exist.

Case 5) x approaches $+\infty$ or $-\infty$ in a $\dfrac{\text{polynomial}}{\text{polynomial}}$. Divide the top and bottom by the

highest power of x in the bottom, to obtain one of these outcomes:

(i) $\dfrac{0}{c}$ (ii) $\dfrac{b}{c}$ (iii) "$\pm\dfrac{\infty}{c}$", where $b, c \neq 0.$

Then the answer is (i) 0 (ii) $\dfrac{b}{c}$ (iii) $+\infty$ or $-\infty$

Example 5) $\lim\limits_{x \to \infty}\left(\dfrac{x^5 + 1}{x^4 + x^3}\right) \overset{\boxed{\text{Divide top and bottom by } x^4.}}{=} \lim\limits_{x \to \infty}\left(\dfrac{x + \dfrac{1}{x^4}}{1 + \dfrac{1}{x}}\right) = \dfrac{\lim\limits_{x \to \infty} x + 0}{1 + 0} = +\infty$

Two for you.

Evaluate: 1) $\lim_{x \to 3^-}\left(\dfrac{3-4x}{x-3}\right)$ 2) $\lim_{x \to -\infty}\left(\dfrac{x^5+1}{x^4+x^3}\right)$

Answers 1) ∞ 2) $-\infty$

Variations on $\lim\limits_{\theta\to 0}\dfrac{\sin\theta}{\theta}=1$

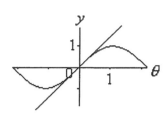

Lots of trigonometric $\dfrac{\text{``}0\text{''}}{0}$ limits are

evaluated with a little clever use of $\lim\limits_{\theta\to 0}\dfrac{\sin\theta}{\theta}=1$.

(By the way, the interpretation of this limit is very intuitive: when θ is small, $\sin\theta$ and θ are nearly equal. Therefore, this ratio approaches 1.)

Example 1) Evaluate the following limits:

(a) $\lim\limits_{\theta\to 0}\dfrac{\theta}{\sin\theta}$ (b) $\lim\limits_{\theta\to 0}\dfrac{\sin 3\theta}{\theta}$ (c) $\lim\limits_{\theta\to 0}\dfrac{\tan 2\theta}{\sin\theta}$ (d) $\lim\limits_{\theta\to 0}\dfrac{1-\cos\theta}{\theta}$

Solution (a) $\lim\limits_{\theta\to 0}\dfrac{\theta}{\sin\theta}=\lim\limits_{\theta\to 0}\dfrac{1}{\left(\dfrac{\sin\theta}{\theta}\right)}=\dfrac{1}{1}=1$

(b) $\lim\limits_{\theta\to 0}\dfrac{\sin 3\theta}{\theta}$ $\boxed{\text{``}3\theta\text{'' is playing the role of ``}\theta\text{'' here!}\atop\text{Put 3 in the bottom and compensate with 3 on top to keep things equal!}}$ $=\lim\limits_{\theta\to 0}\dfrac{3\sin 3\theta}{3\theta}$ $\boxed{3\theta\text{ approaches 0}\atop\text{as }\theta\text{ approaches 0.}}=\lim\limits_{3\theta\to 0}\dfrac{3\sin 3\theta}{3\theta}=3(1)=3$

(c) $\lim\limits_{\theta\to 0}\dfrac{\tan 2\theta}{\sin\theta}$ $\boxed{\text{Use tan}2\theta=\dfrac{\sin 2\theta}{\cos 2\theta}}=\lim\limits_{\theta\to 0}\dfrac{\sin 2\theta}{\sin\theta\cos 2\theta}$ $\boxed{\text{Throw in some}\atop\text{``}\theta\text{'s'' to make the}\atop\text{variables MATCH!}}=\lim\limits_{\theta\to 0}\left(\dfrac{\sin 2\theta}{2\theta}\right)\left(\dfrac{\theta}{\sin\theta}\right)\left(\dfrac{2}{\cos 2\theta}\right)=(1)(1)(2)=2$

(d) $\lim\limits_{\theta\to 0}\dfrac{1-\cos\theta}{\theta}$ $\boxed{\text{Use sin}^2\theta=1-\cos^2\theta.}=\lim\limits_{\theta\to 0}\left(\dfrac{1-\cos\theta}{\theta}\right)\left(\dfrac{1+\cos\theta}{1+\cos\theta}\right)=\lim\limits_{\theta\to 0}\left(\dfrac{1-\cos^2\theta}{\theta(1+\cos\theta)}\right)$

$=\lim\limits_{\theta\to 0}\left(\dfrac{\sin^2\theta}{\theta(1+\cos\theta)}\right)=\lim\limits_{\theta\to 0}\left(\dfrac{\sin\theta}{\theta}\right)\left(\dfrac{\sin\theta}{1+\cos\theta}\right)=(1)\left(\dfrac{0}{2}\right)=0$

157

Two for you.

Evaluate the following limits:

1) $\displaystyle\lim_{\theta \to 0} \frac{\sin 5\theta}{\sin 3\theta}$

2) $\displaystyle\lim_{h \to 0} \frac{\sin(x+h) - \sin x}{h}$ (Hint: use $\sin(x+h) = \sin x \cos h + \cos x \sin h$.)

Answers 1) $\dfrac{5}{3}$ 2) $\cos x$ (Here, you have just shown the derivative of $\sin x$ is $\cos x$!)

Continuity and Discontinuity at a Point

> f is continuous at $x = a$ if $\lim\limits_{x \to a} f(x) = f(a)$.

Continuous functions have no holes. You have to have a limit, you have to have a function value, and they must be equal.

Example 1) For each of the following, explain why the function is discontinuous at the indicated value:

(a) $f(x) = \dfrac{x^2 - 16}{x - 4}$ at $x = 4$ (b) $f(x) = \dfrac{x}{|x|}$ at $x = 0$

(c) $f(x) = \begin{cases} x^2, & \text{if } x \neq 0 \\ 2, & \text{if } x = 0 \end{cases}$ at $x = 0$ (d) $f(x) = [[x]]$ at $x = 2$

Solution (a) $\lim\limits_{x \to 4} f(x) = \lim\limits_{x \to 4} \dfrac{\cancel{(x-4)}(x+4)}{\cancel{x-4}} = 8$, so the limit exists. But $f(4)$ does not exist.

(b) $f(x) = \dfrac{x}{|x|} = \begin{cases} -x, & \text{if } x < 0 \\ x, & \text{if } x > 0 \end{cases}$ $\therefore$ $\lim\limits_{x \to 0^-} f(x) = \lim\limits_{x \to 0^-}(-1) = -1$ and $\lim\limits_{x \to 0^+} f(x) = \lim\limits_{x \to 0^+} 1 = 1$

and so $\lim\limits_{x \to 0} f(x)$ does not exist.

(c) $\lim\limits_{x \to 0} f(x) = \lim\limits_{x \to 0} x^2 = 0$ so the limit exists. But $f(0) = 2 \neq \lim\limits_{x \to 0} f(x)$.

(d) $\lim\limits_{x \to 2^-} f(x) \overset{\boxed{\text{For } 1 \leq x < 2, f(x) = 1.}}{=} \lim\limits_{x \to 2^-}(1) = 1$ and $\lim\limits_{x \to 2^+} f(x) = \lim\limits_{x \to 2^+} 2 = 2$, and so $\lim\limits_{x \to 2} f(x)$ does not exist.

Note that $f(2) = 2$ and so f is continuous from the **right**.

Example 2) For the function at the right, give a reason why it is discontinuous at each of a, b, and c. State whether it is continuous from the left or right.

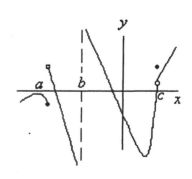

Solution The function is discontinuous at a and b because the limit does not exist. It is discontinuous at c because the limit and value are unequal. The function is continuous from the **left** at a but from neither side at b and c.

Two for you.

For each of the following, give a reason why the function is not continuous at the indicated value:

1) $f(x) = \dfrac{1}{x-3}$ at $x = 3$

2) $f(x) = \begin{cases} x+1, & \text{if } x < -2 \\ 2, & \text{if } x = -2 \\ x^2 - 5, & \text{if } x > -2 \end{cases}$ at $x = -2$

Answers 1) $\lim\limits_{x \to 3} f(x)$ does not exist **OR** $f(3)$ does not exist (either answer).

2) $\lim\limits_{x \to -2} f(x) = -1 \neq f(-2)$

160

Continuous Functions (Intervals of Continuity)

Continuous functions have no holes. When you draw a continuous function, your pen never leaves the paper. When is a function **not continuous**? Basically, you have to check **domain restrictions**, **division by 0**, and **"branches"** (which we look at geometrically in Example 2 and using function definitions on the next page!) When you add, subtract, and multiply continuous functions, you get continuous functions. Same for division, as long as you don't have 0 in the bottom!

Example 1)(a) Based on your mathematical experience, which of the following are continuous polynomials, trigonometric functions, $y = a^x$, $y = \log_a x$?

(b) If f and g are continuous, discuss the continuity of $f + g$, $f - g$, fg, f/g.

Solution (a) Polynomials, $y = \sin x$, $y = \cos x$, and $y = a^x$ are continuous for $x \in \mathbb{R}$.

$y = \log_a x$ is continuous for $x > 0$.

$y = \tan x$ and $y = \sec x$ are continuous for $x \neq \pi/2 + 2k\pi$, where $k \in \mathbb{Z}$.

$y = \cot x$ and $y = \csc x$ are continuous for $x \neq \pi + 2k\pi$, where $k \in \mathbb{Z}$.

(b) $f + g$, $f - g$, and fg are all continuous whenever f and g are.

f/g is continuous as long as f and g are and $g(x) \neq 0$.

Example 2) State the intervals of continuity

for this function →

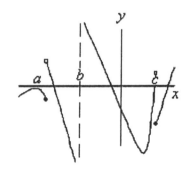

Solution The function is continuous on the intervals $(-\infty, a]$, (a, b), (b, c), and $[c, \infty)$.

Note that the function is continuous from the **left** at $x = a$ and from the **right** at $x = c$.

Example 3) State the intervals on which the following are continuous:

(a) $f(x) = \sqrt{x - 3}$ (b) $g(x) = \ln(|x|)$ (c) $h(x) = \dfrac{x^{1/4}}{(x - 5)(x + 3)}$

Solution (a) Since $x - 3 \geq 0$, therefore $x \geq 3$ and so f is continuous on the interval $[3, \infty)$.

(b) Since $|x|$ is always positive, we only have to exclude $x = 0$ and so g is continuous on $(-\infty, 0)$ and $(0, \infty)$.

(c) $x \neq 5$ or -3. But $x^{1/4}$ is only defined for $x \geq 0$, so -3 is already gone. Therefore h is continuous on $[0, 5)$ and $(5, \infty)$.

161

Two for you.

State the intervals of continuity for f and g: 1) $f(x) = \dfrac{\sqrt{3x-9}}{x-5}$

2) $g(x) = [[x]]$, where $-2 \le x \le 2$ (g is "the greatest integer less than or equal to x" function.)

Answers 1) $[3,5)$, $(5,\infty)$ 2) $[-2,-1)$, $[-1,0)$, $[0,1)$, $[1, 2)$

Continuity and Branch Functions

There are two kinds of branch functions: explicit and implicit. Apart from the usual continuity concerns (division by 0, positive domain for logs, etc.) you have to be careful **at the values of x where the function branches.** Remember:

$\boxed{f \text{ is continuous at } x = a \text{ if } \lim_{x \to a} f(x) = f(a).}$

Example 1) Discuss the continuity of

the function $f(x) = \begin{cases} x, \text{ if } x < 0 \\ 2, \text{ if } x = 0 \\ x^2, \text{ if } 0 < x < 2 \\ x + 2, \text{ if } x \geq 2 \end{cases}$

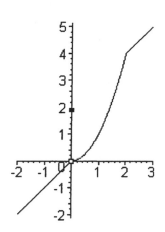

Solution The only possible problems
are at $x = 0$ and $x = 2$.

$\lim_{x \to 0^-} f(x) = \lim_{x \to 0^-} x = 0 \qquad \lim_{x \to 0^+} f(x) = \lim_{x \to 0^+} x^2 = 0$

Since the left- and right-hand limits are equal, $\lim_{x \to 0} f(x) = 0$.

But $f(0) = 2$ and so f **is not** continuous at $x = 0$.

$\lim_{x \to 2^-} f(x) = \lim_{x \to 2^-} x^2 = 4 \qquad \lim_{x \to 2^+} f(x) = \lim_{x \to 2^+} (x + 2) = 4$. Since the left- and right-hand limits are equal

$\lim_{x \to 2} f(x) = 4$. Also, $f(2) = 4$ and so f **is** continuous at $x = 2$. f is continuous on the intervals

$(-\infty, 0)$ and $(0, \infty)$.

Example 2) Discuss the continuity of the function $g(x) = \dfrac{|x|}{x}$.

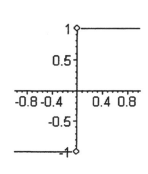

Solution The only possible problem here is at $x = 0$.
Writing g explicitly as a branch function,

we have $g(x) = \begin{cases} \dfrac{-x}{x}, \text{ if } x < 0 \\ \dfrac{x}{x}, \text{ if } x > 0 \end{cases} = \begin{cases} -1, \text{ if } x < 0 \\ 1, \text{ if } x > 0 \end{cases}$

$\lim_{x \to 0^-} g(x) = \lim_{x \to 0^-} (-1) = -1$ and $\lim_{x \to 0^+} g(x) = \lim_{x \to 0^+} = 1$

so $\lim_{x \to 0} g(x)$ does not exist. Therefore, g is **not continuous**

at $x = 0$ and so g is continuous on the intervals $(-\infty, 0)$ and $(0, \infty)$.

Two for you.

State the intervals of continuity for f and g: 1) $f(x)=\begin{cases} \sin x, & \text{if } x < \pi \\ \cos x, & \text{if } x \geq \pi \end{cases}$ 2) $g(x)=\dfrac{x+3}{|x+3|}$

Answers 1) $(-\infty, \pi)$, $[\pi, \infty)$ 2) $(-\infty, -3)$, $(-3, \infty)$

Essential versus Removable Discontinuities

This is easy!

The discontinuity is **removable** if it is a "hole" in the function. In this case, there is a limit and there is a function value, and they are different **or** there is no function value.

The discontinuity is **essential** if it is **any other type of discontinuity**. In this case, there is definitely **NO** limit, though this can happen for a variety of reasons.

Example 1) Each of the following functions has a discontinuity at one point. Classify it as either essential or removable. If essential, give the reason. If removable, tell how to define the function to make it continuous at the point.

(a) $f(x) = \begin{cases} \sin x, \text{if } x < 0 \\ 1, \text{if } x = 0 \\ x, \text{if } x > 0 \end{cases}$
(b) $g(x) = \dfrac{x^2 - 9}{x - 3}$
(c) $h(x) = \dfrac{1}{x^2}$

(d) $j(x) = \begin{cases} x, \text{if } x < 0 \\ 1, \text{if } x = 0 \\ x + 1, \text{if } x > 0 \end{cases}$
(e) $k(x) = \sin\left(\dfrac{1}{x}\right)$

Solution (a) Removable. Define $f(0) = 0$.

(b) Removable. The problem is at $x = 3$: $\lim\limits_{x \to 3} g(x) = \lim\limits_{x \to 3} \dfrac{(x - 3)(x + 3)}{x - 3} = 6$, but $g(3)$ is not defined

Define $g(3) = 6$.

(c) Essential. $\lim\limits_{x \to 0} h(x) = \infty$

(d) Essential. $\lim\limits_{x \to 0^-} j(x) = \lim\limits_{x \to 0^-} x = 0$ while $\lim\limits_{x \to 0^+} j(x) = \lim\limits_{x \to 0^+} (x + 1) = 1$
and so $\lim\limits_{x \to 0} j(x)$ does not exist.

(e) Essential. $\lim\limits_{x \to 0} k(x)$ does not exist. In fact, $\sin\left(\dfrac{1}{x}\right)$ **oscillates** between -1 and 1 **faster**

and **faster** as x gets closer and closer to 0. It is actually quite "dizzying" and fun to watch!

Two for you.

Each function has a single discontinuity. State the problem x value and whether the discontinuity is essential or removable. If essential, state why. If removable, define the function to make it continuo...

1) $f(x) = \dfrac{|x-5|}{x-5}$ 2) $g(x) = \dfrac{\sin\theta}{\theta}$

Answers 1) Essential at $x = 5$ since $\lim\limits_{x\to 5} f(x)$ does not exist.

2) Removable at $\theta = 0$. Define $g(0) = 1$.

Finding the Derivative from the Definition

f is differentiable at $x = a$ if $\displaystyle\lim_{h \to 0} \frac{f(a+h) - f(a)}{h} = f(a)$.

The alternate form of this definition is

f is differentiable at $x = a$ if $\displaystyle\lim_{x \to a} \frac{f(x) - f(a)}{x - a} = f(a)$.

Forgive me, but I am about to be very wordy!

The second definition is, **believe me**, the same as the first. Many of you find this hard to believe. Trust me. The **variable** in the second formulation is $x = (a + h)$ and so $(x - a) = h$. As h approaches 0, x approaches a. So, in the first definition, replace:

$h \to 0$ with $x \to a$ h with $(x - a)$ $(a + h)$ with x.

Lo and behold but not by magic, you have the second definition. Why do students get these confused? I think it is because teachers ask you to find $f'(\boxed{x})$, rather than $f'(\boxed{a})$, from the definition when you first study derivatives. This is okay in the first definition where x takes on the role of a. **BUT** if you try to use the second definition, you would be forcing x to be both $(a + h)$ and a at the same time!

Example 1) Find the derivative for the function $f(x) = x^2 + x + 1$ at the point $(a, f(a))$
(a) using the first definition   (b) using the second definition.

Solution (a) $f'(a) = \displaystyle\lim_{h \to 0} \frac{f(a+h) - f(a)}{h} = \lim_{h \to 0} \frac{(a+h)^2 + (a+h) + 1 - (a^2 + a + 1)}{h}$

$= \displaystyle\lim_{h \to 0} \frac{a^2 + 2ah + h^2 + a + h + 1 - a^2 - a - 1}{h} = \lim_{h \to 0} \frac{2ah + h^2 + h}{h} = \lim_{h \to 0} \frac{\cancel{h}(2a + h + 1)}{\cancel{h}} = 2a + 1$

(b) $f'(a) = \displaystyle\lim_{x \to a} \frac{f(x) - f(a)}{x - a} = \lim_{x \to a} \frac{x^2 + x + 1 - (a^2 + a + 1)}{x - a}$

$= \displaystyle\lim_{x \to a} \frac{x - a^2 + x - a}{x - a} = \lim_{x \to a} \frac{(x - a)(x + a + 1)}{x - a} = \lim_{x \to a} \frac{\cancel{(x - a)}(x + a + 1)}{\cancel{x - a}} = 2a + 1$

Two for you.

Set up the first step to find $f'(a)$ for $f(x) = \dfrac{1}{\sqrt{x}}$

1) using the first definition   2) using the second definition.

Answers 1) $f'(a) = \lim\limits_{h \to 0} \dfrac{\dfrac{1}{\sqrt{a+h}} - \dfrac{1}{\sqrt{a}}}{h}$ 2) $f'(a) = \lim\limits_{x \to a} \dfrac{\dfrac{1}{\sqrt{x}} - \dfrac{1}{\sqrt{a}}}{x - a}$

Differentiable Functions (Intervals of Differentiability)

When is a **continuous** function **not differentiable**?

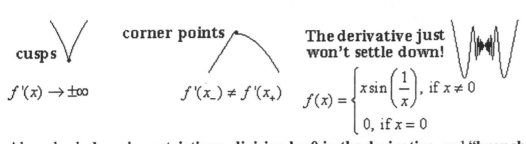

cusps $f'(x) \to \pm\infty$

corner points $f'(x_-) \neq f'(x_+)$

The derivative just won't settle down! $f(x) = \begin{cases} x\sin\left(\dfrac{1}{x}\right), & \text{if } x \neq 0 \\ 0, & \text{if } x = 0 \end{cases}$

Also, check **domain restrictions, division by 0 in the derivative**, and "**branches**". When you add, subtract and multiply differentiable functions, you get differentiable functions. Same for division, as long as you don't have 0 in the bottom!

Example 1) Based on your mathematical experience, which of the following are differentiable: polynomials, trigonometric functions, $y = a^x$, $y = \log_a x$, $y = |x|$?

Solution Polynomials, $y = \sin x$, $y = \cos x$, and $y = a^x$ are differentiable for $x \in \mathbb{R}$.

$y = \log_a x$ is differentiable for $x > 0$.

$y = \tan x$ and $y = \sec x$ are differentiable for $x \neq \pi/2 + 2k\pi$, where $k \in \mathbb{Z}$.

$y = \cot x$ and $y = \csc x$ are differentiable for $x \neq \pi + 2k\pi$, where $k \in \mathbb{Z}$.

$y = |x|$ has a corner point at $x = 0$ $(f'(0_-) \neq f'(0_+))$ and so the function is differentiable for $x \neq 0$

Example 2) State the intervals of differentiability for this function →

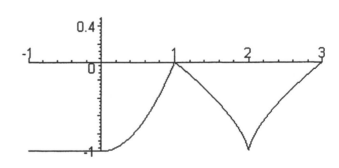

Solution The function is differentiable on the intervals $[-1, 1]$, $[1, 2)$, and $(2, 3]$. Note that $x = 1$ is a corner point and $x = 2$ is a cusp point.

Example 3) State the intervals on which the following are differentiable.

(a) $f(x) = \sqrt{x - 3}$ (b) $h(x) = \dfrac{x^{1/4}}{(x - 5)(x + 3)}$

Solution (a) Since $x - 3 \geq 0$, therefore $x \geq 3$ and so f is differentiable on the interval $[3, \infty)$.

(c) $x \neq 5$ or -3. But $x^{1/4}$ is only defined for $x \geq 0$, so -3 is already gone! Therefore, h is differentiable on the intervals $[0, 5)$ and $(5, \infty)$.

Two for you.

State the intervals of differentiability for 1) $f(x) = \dfrac{\sqrt{3x-9}}{x-5}$ 2) $g(x) = [[x]]$, if $-2 \le x \le 2$

Answers 1) $[3, 5), (5, \infty)$ 2) $[-2, -1), [-1, 0), [0, 1), [1, 2)$

Differentiability and Branch Functions

There are two kinds of branch functions: explicit and implicit. Apart from the usual differentiability concerns (division by 0 in the derivative, positive domain for logs, etc.) you have to be careful **at the values of x where the function branches. Also, we will deal here only with functions that are continuous on their domains.**

Example 1) Discuss the differentiability of

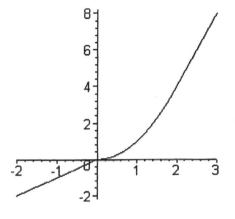

the function $f(x) = \begin{cases} x, & \text{if } x < 0 \\ x^2, & \text{if } 0 \leq x < 2 \\ 4x - 4, & \text{if } x \geq 2 \end{cases}$

Solution The only possible problems are at $x = 0$ and $x = 2$.

$$\lim_{x \to 0^-} f'(x) = \lim_{x \to 0^-} 1 = 1 \quad \lim_{x \to 0^+} f'(x) = \lim_{x \to 0^+} 2x = 0$$

Since the left- and right-hand limits are unequal, $f'(0)$ does not exist.

Note f is differentiable from both the left and the right at $x = 0$.

$$\lim_{x \to 2^-} f'(x) = \lim_{x \to 2^-} 2x = 4 \quad \lim_{x \to 2^+} f'(x) = \lim_{x \to 2^+} 4 = 4. \text{ Therefore,}$$

$f'(2) = 4$ and so f is differentiable for $x \neq 0$.

Note : If a function is discontinuous at a value of x, it can only be continuous from the left or the right or neither, but not both. As this example shows, a continuous function may not be differentiable at a value of x yet still be differentiable from the left and the right.

Example 2) Discuss the differentiability of the function $g(x) = |x^3|$.

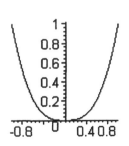

Solution The only possible problem here is at $x = 0$.

Writing g explicity as a branch function,

we have $g(x) = \begin{cases} -x^3, & \text{if } x < 0 \\ x^3, & \text{if } x \geq 0 \end{cases} = \begin{cases} -1, & \text{if } x < 0 \\ 1, & \text{if } x \geq 0 \end{cases}$

$$\lim_{x \to 0^-} g'(x) = \lim_{x \to 0^-} (-3x^2) = 0 \quad \text{and} \quad \lim_{x \to 0^+} g(x) = \lim_{x \to 0^+} 3x^2 = 0$$

so $g'(0) = 0$ and g is differentiable for $x \in \mathbb{R}$.

Two for you.

State the intervals of differentiability for 1) $f(x)=\begin{cases} \sin x, \text{if } x < \pi \\ \cos x - x, \text{if } x \geq \pi \end{cases}$ 2) $g(x)=|x+3$

Answers 1) $(-\infty,\infty)$ 2) $(-\infty,-3)$ and $(-3,\infty)$

Critical Numbers

A **critical number** of a function $y = f(x)$ is a number c ⎢in the domain of the function⎥ where (i) $f'(c) = 0$ or (ii) $f'(c)$ does not exist or (iii) $(c, f(c))$ is an end point. We all know what 0 derivatives and end points look like. If $f(x)$ is continuous at c but $f'(c)$ does not exist, then one of three things must be happening:

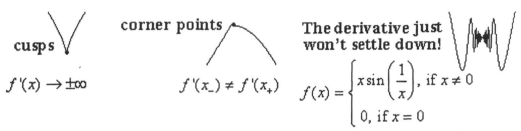

cusps
$f'(x) \to \pm\infty$

corner points
$f'(x_-) \neq f'(x_+)$

The derivative just won't settle down!
$$f(x) = \begin{cases} x\sin\left(\dfrac{1}{x}\right), & \text{if } x \neq 0 \\ 0, & \text{if } x = 0 \end{cases}$$

A maximum or minimum point of a continuous function f **must occur** at a critical number of f!

⎢**The reverse is FALSE! Lots of critical points are neither maxima nor minima!**⎥

Example 1) Find all critical numbers of the function $f(x) = 3x^4 + 2x^3$, $-2 \leq x \leq 2$ and classify the corresponding critical points.

Solution $f'(x) = 12x^3 + 6x^2 = 12x^2\left(x + \dfrac{1}{2}\right)$

Both f and f' have domain $[-2, 2]$. The critical numbers are -2, $-\dfrac{1}{2}$, 0, and 2.

$f'(x)$

```
    ├───────────┼────┼──────────────┤
    -2     −    -1/2 + 0      +      2
    │      ╲     │ ╱ │      ╱        │
```

$(-2, -32)$ is an end point maximum; $(-1/2, -1/16)$ is a minimum where $f' = 0$; $(0, 0)$ is neither a max nor a min. (In fact, it is a point of inflection.); $(2, 64)$ is an end point maximum

Example 2) Find all critical numbers of the function $f(x) = 5x^{2/3} - x^{5/3}$.

Solution $f'(x) = \dfrac{10}{3}x^{1/3} - \dfrac{5}{3}x^{2/3} = -\dfrac{5}{3}\left(\dfrac{x - 2}{x^{1/3}}\right)$

The critical numbers are 0 and 2. $\Big($ Note that $0 \in \text{domain}(f)$ but not $\text{domain}(f')$.$\Big)$

Two for you.

Find the critical numbers for each of the following functions:

1) $f(x) = \dfrac{x^2}{x^2 - 4}$

2) $g(x) = (x+1)^{1/3}(x-1)^{2/3}$ $\left(\text{Hint: } f'(x) = \dfrac{x + 1/3}{(x+1)^{2/3}(x-1)^{1/3}} \right)$

Answers 1) $x = 0$ 2) $x = -\dfrac{1}{3}$, $x = -1$, $x = 1$

Max and Min Points from the First Derivative

There are **four** ways a **continuous** function can have a maximum or minimum point:

1) $f'(x) = 0$ 2) $f'(x) \rightarrow \pm\infty$ 3) $f'(x_-) \neq f'(x_+)$ 4) an end point

(a cusp)   (a corner point)

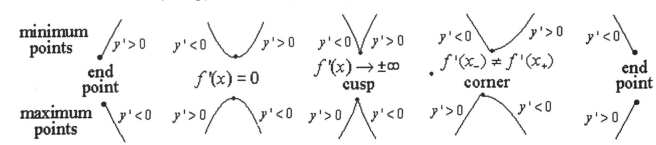

What all minimums have in common is that the function decreases ($y' < 0$) and then increases ($y' > 0$).

What all maximums have in common is that the function increases ($y' > 0$) and then decreases ($y' < 0$)

(For end points, "half" of each statement applies!)

Example 1) Given $f'(x) = \dfrac{(x-2)^5(x+2)^3}{(x-1)^{1/3}}$, $-3 \leq x \leq 3$, identify the values of x where $f(x)$ has

maximum and minimum points. Classify these extremes as one of the four types.
(You may assume that $f(x)$ is defined for $x \in [-3,3]$, including $x = 1$.)

Solution Analyze the sign of $f'(x)$ on $[-3,3]$. The only significant values are $x = -3, -2, 1, 2,$ and 3

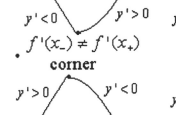

$f'(x)$ line analysis:

$$\begin{array}{ccccc}
 & (-)(-)(-) & (+)(-)(-) & (+)(+)(-) & (+)(+)(+) \\
-3 & - & -2 & + & 1 & - & 2 & + & 3 \\
 & \text{dec} \searrow & & \text{inc} \nearrow & & \text{dec} \searrow & & \text{inc} \nearrow &
\end{array}$$

$x = -3$:	end point, $(-3, f(-3))$ is a maximum point.
$x = -2$:	$f'(-2) = 0$, $(-2, f(-2))$ is a minimum point.
$x = 1$:	$f'(1) \rightarrow \pm\infty$, cusp, $(1, f(1))$ is a maximum point.
$x = 2$:	$f'(2) = 0$, $(2, f(2))$ is a minimum point.
$x = 3$:	end point, $(3, f(3))$ is a maximum point.

Two for you.

For the following, identify the values of x where the function has maxima or minima and classify each extreme:

1) $f'(x) = -\dfrac{x-2}{x^{1/3}}$, where $x \geq -1$ (Hint: be careful with that extra "–"!)

2) $g'(x) = \begin{cases} -x, & \text{if } x < 0 \\ x^2, & \text{if } x \geq 0 \end{cases}$

Answers 1) $x = -1$: end point, maximum; $x = 0$, cusp, minimum; $x = 2$, $f'(x) = 0$, maximum
2) $x = 0$, corner point.

Graphing and Interpreting y vs y' vs y''

Here is a function that illustrates:
$y' > 0$ and $y'' > 0$: y is increasing and concave up.
$y' > 0$ and $y'' < 0$: y is increasing and concave down.
$y' < 0$ and $y'' > 0$: y is decreasing and concave up.
$y' < 0$ and $y'' < 0$: y is decreasing and concave down.

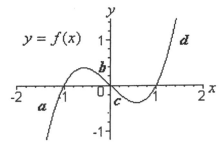

Example 1) Insert the letters **a, b, c,** and **d** with the correct choice of y' and y'' in the above graph

$y' > 0$, $y'' > 0$ ☐ $y' > 0$, $y'' < 0$ ☐ $y' < 0$, $y'' > 0$ ☐ $y' < 0$, $y'' < 0$ ☐

Solution $y' > 0$, $y'' > 0$ $\boxed{d}$ $y' > 0$, $y'' < 0$ $\boxed{a}$ $y' < 0$, $y'' > 0$ $\boxed{c}$ $y' < 0$, $y'' < 0$ $\boxed{b}$

Example 2)(a) By estimating the slopes of the tangents to $y = f(x)$, sketch the graph of $y' = f'(x)$.
(b) By estimating the slopes of the tangents to $y' = f'(x)$, sketch the graph of $y'' = f''(x)$.

Solution   (a)
m = slope

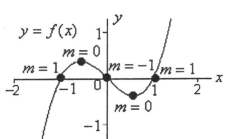

(b)

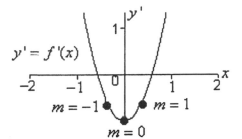

Two for you.

Match the shapes with the appropriate pair of derivatives as in Example 1:

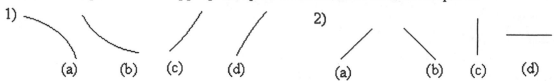

1)
(a)   (b)   (c)   (d)

2)
(a)   (b)   (c)   (d)

Answers 1)(a) $y' < 0,\ y'' < 0$ (b) $y' < 0,\ y'' > 0$ (c) $y' > 0,\ y'' > 0$ (d) $y' > 0,\ y'' < 0$

2)(a) $y' > 0,\ y'' = 0$ (b) $y' < 0,\ y'' = 0$ (c) $y',\ y''$ undefined   (d) $y' = y'' = 0$

Estimating Using the Differential

Let $y = f(x)$. Then $\dfrac{dy}{dx} = f'(x)$, which gives the slope of the tangent at the point $(x, f(x))$.

If we treat $dx\,(\neq 0)$ as the run and dy as the rise so that $\dfrac{dy}{dx} = f'(x) =$ slope of the tangent,

then we can rewrite this as $\boxed{dy = f'(x)\,dx, \text{ in words, } \textbf{the rise} = \textbf{slope times the run.}}$

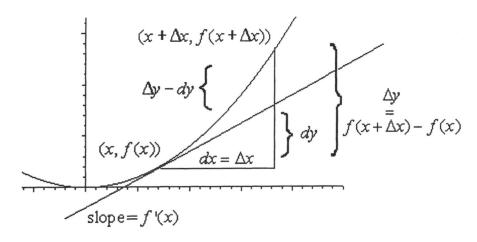

The key: when dx is small, dy **is very nearly equal** Δy (← **Look at the picture!**), so

$$\boxed{\text{the new } y = f(x + \Delta x) = \text{old } y + \Delta y \overset{\text{when } dx \text{ is small!}}{\doteq} \text{old } y + dy = f(x) + dx}$$

Example 1) Estimate $\sqrt{4.02}$ using the differential.

Solution We need to choose an appropriate function $f(x)$, a value of x close to 4.02 at which we can **easily** evaluate the function and a **small dx** value.

Let $f(x) = \sqrt{x}$, $x = 4$, and $dx = 0.02$. Then $\dfrac{dy}{dx} = f'(x) = \dfrac{1}{2}x^{-1/2} = \dfrac{1}{2\sqrt{x}}$.

$\therefore dy = f'(x)\,dx = \dfrac{dx}{2\sqrt{x}}$. When $x = 4$ and $dx = 0.02$, we have $dy = \dfrac{0.02}{2\sqrt{4}} = \dfrac{1}{200} = 0.005$

$\therefore \sqrt{4.02} \overset{\text{exactly}}{=} \sqrt{4} + \Delta y \doteq \sqrt{4} + dy = 2.005.$

Compare $\sqrt{4.02} \overset{\text{calculator}}{\doteq} 2.00499376558$. Pretty good!

179

Two for you.

1) Estimate $\sqrt{3.98}$ using the differential. (Hint: this is just like the example above but use $dx = -0.0$

2) Give the best choice for $f(x)$, x, and dx in order to estimate $\dfrac{1}{26^{1/3}}$ using the differential.

Answers 1) 1.995   2) $f(x) = \dfrac{1}{x^{1/3}}$, $x = 27$, $dx = -1$

Rolle's Theorem

Suppose $y = f(x)$ is

(i) continuous on $[a, b]$

(ii) differentiable on (a, b) and

(iii) $f(a) = f(b)$.

Then there is **at least one** value

$c \in (a,b)$ such that $f'(c) = 0$.

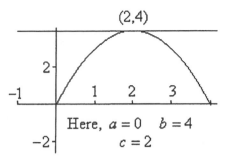

Here, $a = 0$ $b = 4$

$c = 2$

So, basically, Rolle's says that if a continuous function goes, for example, up and then turns around and comes back down to its starting value, then it must have a maximum, say at $x = c$. Since the function is differentiable, this max can't be a corner point nor a cusp. Since it certainly isn't an end point, the only other choice is $f'(c) = 0$.

Example 1) Verify Rolle's Theorem for the function $f(x) = x^2 - 3$, with $a = -1$ and $b = 1$.

Solution f is certainly differentiable and continuous on $(-1,1)$ and $[-1,1]$, respectively. Also, $f(-1) = f(1) = -2$. We should, according to Rolle's Theorem, be able to find c strictly between -1 and 1 satisfying $f'(c) = 0$. Since $f'(x) = 2x$, solving $2c = 0$, we find $c = 0 \in (-1,1)$.

Example 2) Sketch the graph of a function showing

(a) how the conclusion of Rolle's Theorem **can fail** if

    (i) f is not continuous at a

    (ii) f is not differentiable for some $x \in (a,b)$ and

(b) how Rolle's Theorem **will succeed** even if f is not differentiable at $x = a$. (This illustrates that f doesn't have to be differentiable at $x = a$ for Rolle's Theorem!)

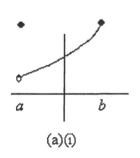

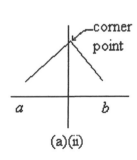

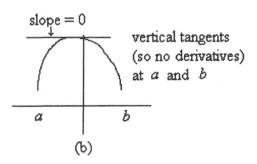

(a)(i)   (a)(ii)   (b)

(no zero slope)   (no zero slope)   (There **is**--and must be--a zero slope!)

Two for you.

1) Verify Rolle's Theorem for the function $f(x) = x^3 - x + 3$, with $a = 0$ and $b = 1$.

2) Sketch a function that illustrates how the conclusion of Rolle's Theorem **can fail**, that is, there will be no 0 derivative, if f is not continuous for some $x \in (a,b)$.

Answers 1) $f'\left(\dfrac{1}{\sqrt{3}}\right) = 0$ and $\dfrac{1}{\sqrt{3}} \in (0,1)$, as required. 2)

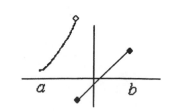

The Mean Value Theorem

Suppose $y = f(x)$ is
(i) continuous on $[a,b]$ and
(ii) differentiable on (a,b).
Then there is **at least one** value

$c \in (a,b)$ such that $f'(c) = \dfrac{f(b)-f(a)}{b-a}$.

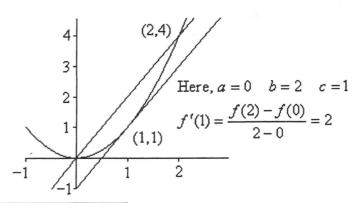

Here, $a = 0$ $b = 2$ $c = 1$

$f'(1) = \dfrac{f(2)-f(0)}{2-0} = 2$

Note : when $f(b) = f(a)$, the MVT becomes Rolle's Theorem.

Personal note : to me, the MVT always seems to be Rolle's Theorem after a few too many drinks!

Example 1) Verify The Mean Value Theorem for the function $f(x) = x^2 - 1$, with $a = -1$ and $b = 2$.

Solution Since f is a polynomial, it is differentiable and continuous on $(-1,2)$ and $[-1,2]$, respectively. According to The Mean Value Theorem, we should be able to find c strictly between

-1 and 2 satisfying $f'(c) = \dfrac{f(2)-f(-1)}{2-(-1)} = \dfrac{3-0}{3} = 1.$

Since $f'(x) = 2x$, solving $2c = 1$, we find $c = \dfrac{1}{2} \in (-1,2)$.

Example 2) Sketch the graph of a function showing
(a) how the conclusion of The Mean Value Theorem **can fail** if
   (i) f is not continuous at a
   (ii) f is not differentiable for some $x \in (a,b)$ and
(b) how The Mean Value Theorem **will succeed** even if f is not differentiable at $x = a$. (This illustrates that f doesn't have to be differentiable at $x = a$ for The Mean Value Theorem!)

Solution

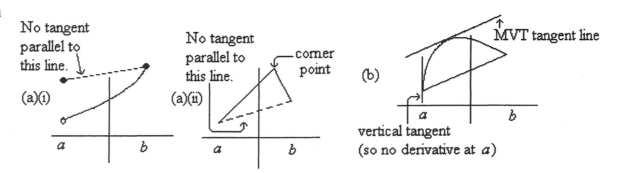

No tangent parallel to this line.

(a)(i)

No tangent parallel to this line.

corner point

(a)(ii)

(b)

MVT tangent line

vertical tangent (so no derivative at a)

183

Two for you.

1) Verify The Mean Value Theorem for the function $f(x) = x^3 + 3$, with $a = -1$ and $b = 2$.

2) Sketch a function **not continuous** at some $x \in (a, b)$ for which the conclusion of the MVT **fails**.

Answers 1) $f'(1) = 3$ and $1 \in (-1, 2)$, as required.   2)

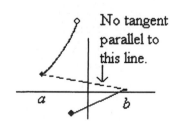

No tangent parallel to this line.

Derivatives: The Product Rule

Think of **First** and **Second**:

$$\frac{d}{dx}(FS) = FS' + SF' \text{! Note: the order is NOT important!}$$

Example 1) Find $\frac{dy}{dx}$ if $y = (x^3 + x)(\sin x - 4)$.

Solution $\frac{dy}{dx} = (x^3 + x)\cos x + (\sin x - 4)(3x^2 + 1) = x^3 \cos x + x\cos x + 3x^2 \sin x - 12x^2 + \sin x - 4$

Here is the product rule combined with the **chain rule** (reviewed on the next page!)

Example 2) Find $\frac{dy}{dx}$ if $y = x\left(e^x + \ln x\right)^5$.

Solution $\frac{dy}{dx} = x\left(5\left(e^x + \ln x\right)^4\left(e^x + \frac{1}{x}\right)\right) + \left(e^x + \ln x\right)^5 (1) = \left(e^x + \ln x\right)^4\left(5x\left(e^x + \frac{1}{x}\right) + \left(e^x + \ln x\right)\right)$

$= \left(e^x + \ln x\right)^4\left(5xe^x + 5 + e^x + \ln x\right)$

Example 3) Find the formula for the derivative of the product of three functions,

that is, find $\frac{dy}{dx}$ when $y = f(x)g(x)h(x)$.

Solution Group two of the functions together and use the basic product rule.
Writing $y = \left(f(x)g(x)\right)h(x)$,

$\frac{dy}{dx} = \left(f(x)g(x)\right)h'(x) + h(x)\left(f(x)g(x)\right)'$

$= \left(f(x)g(x)\right)h'(x) + h(x)\left(f(x)g'(x) + g(x)f'(x)\right) = f(x)g(x)h'(x) + f(x)h(x)g'(x) + g(x)h(x)f'(x)$

Three for you.

1) Find $\dfrac{dy}{dx}$ for each of the following: (a) $y = x^4(x^5 + 10)^6$ (b) $y = (5x + 4)(\sin x)(\ln x)$

2) Write the formula for $\dfrac{dy}{dx}$ if $y = f_1 f_2 f_3 f_4$.

Answers 1)(a) $2x^3(x^5 + 10)^5(17x^5 + 10)$

(b) $\dfrac{(5x + 4)(\sin x)}{x} + (5x + 4)(\ln x)(\cos x) + 5(\sin x)(\ln x)$

2) $f_1 f_2 f_3 f'_4 + f_1 f_2 f'_3 f_4 + f_1 f'_2 f_3 f_4 + f'_1 f_2 f_3 f_4$

Derivatives: The Chain Rule

Example 1) Find $\dfrac{dy}{dx}$ if $y = (\sin x + x)^3$.

Solution Think of this as $y = (inside)^3$.

The chain rule tells us to take the derivative of the **outside** and multiply by the derivative of the **inside**. Keep taking the derivative until you "get to the derivative with respect to x." So…

$$\frac{dy}{dx} = \frac{d\left(inside^3\right)}{d\,(inside)}\frac{d(inside)}{dx} = 3(inside)^2\frac{d(inside)}{dx} = 3(\sin x + x)^2(\cos x + 1)$$

Now, repeating myself in my old age, we keep going until we finally get down to x. If you can do this next example, you are a chain rule pro!

Example 2) Find $\dfrac{dy}{dx}$ if $y = \sqrt{x + \sqrt{x + \sqrt{x}}}$. Here we are going to have go inside **once! twice!! three!!! times!**

Solution I find it much easier to handle an example like this if I rewrite it as

$$y = \sqrt{x + \sqrt{x + \sqrt{x}}} = \left(x + \left(x + x^{1/2}\right)^{1/2}\right)^{1/2}.$$

We are going to take the derivative of the outside and then multiply by the derivative of the **FIRST** inside and then go inside again and, for the final $x^{1/2}$, a third time!

$$\frac{dy}{dx} = \frac{1}{2}\left(\boxed{\text{Now take this inside derivative}}\atop{x + \left(x + x^{1/2}\right)^{1/2}}\right)^{-1/2}\left(1 + \frac{1}{2}\left(\boxed{\text{Now take this inside derivative!}}\atop{x + x^{1/2}}\right)^{-1/2}\left(\boxed{\text{We are down to }x,\text{ finally!}}\atop{1 + \frac{1}{2}x^{-1/2}}\right)\right)$$

$$= \frac{1}{2\sqrt{x + \sqrt{x + \sqrt{x}}}}\left(1 + \frac{1}{2\sqrt{x + \sqrt{x}}}\left(1 + \frac{1}{2\sqrt{x}}\right)\right)$$

187

Two for you.

Find $\dfrac{dy}{dx}$ for each of the following: 1) $y = \sin(x^3 + 3x^2 + 1)$

2) $y = \dfrac{1}{(2x-1)^{3/2}}$ $\left(\text{Hint: first rewrite this as } y = (2x-1)^{-3/2}.\right)$

Answers 1) $(3x^2 + 6x)\cos(x^3 + 3x^2 + 1)$ 2) $-3(2x-1)^{-5/2}$

optional back of book answer $= \dfrac{-3}{(2x-1)^{5/2}}$

188

Derivatives: The Quotient Rule

Think of **Top** (numerator) and **Bottom** (denominator):

$$\frac{d}{dx}\left(\frac{T}{B}\right)=\frac{BT'-TB'}{B^2}.$$ **Note** the "−". **Note** the order. **Note, no** "B'", **that is, no** $\frac{dB}{dx}$, in the bottom!

Example 1) Find $\frac{dy}{dx}$ if $y=\frac{5x^2}{x^2-\sin x}$.

Solution $\frac{dy}{dx}$ $\boxed{\text{Keep the constant 5 outside!}}$ $=\frac{5[(x^2-\sin x)(2x)-x^2(2x-\cos x)]}{(x^2-\sin x)^2}$

$=\frac{5[2x^3-2x\sin x-2x^3+x^2\cos x]}{(x^2-\sin x)^2}=\frac{5(x^2\cos x-2x\sin x)}{(x^2-\sin x)^2}$

Here is a common use of the chain and quotient rules together:

Example 2) Find $\frac{dy}{dx}$ if $y=\left(\frac{3x+5}{4x+7}\right)^5$.

Solution $\frac{dy}{dx}=5\left(\frac{3x+5}{4x+7}\right)^4\left(\frac{(4x+7)(3)-(3x+5)(4)}{(4x+7)^2}\right)$

$\boxed{\begin{array}{l}\text{Combine the }(4x+7)\\\text{factors in the denominator.}\end{array}}$ $=5\left(\frac{(3x+5)^4}{(4x+7)^6}\right)(12x+21-12x-20)=\frac{5(3x-5)^4}{(4x+7)^6}$

Alternate Solution Use the product rule with the chain rule: $y=\left(\frac{3x+5}{4x+7}\right)^5=(3x+5)^5(4x+7)^{-5}$

$\frac{dy}{dx}=(3x+5)^5(-5)(4x+7)^{-6}(4)+(4x+7)^{-5}(5)(3x+5)^4(3)$

$\boxed{\text{Take out the common factors.}}$ $=5(3x+5)^4(4x+7)^{-6}\left(-4(3x+5)+3(4x+7)\right)*$

$=5(3x+5)^4(4x+7)^{-6}(-12x-20+12x+21)=\frac{5(3x+5)^4}{(4x+7)^6}$

*Note that 4 was the lower of the two exponents on $(3x+5)$
and -6 was the lower exponent of the two exponents on $(4x+7)$.

Two for you.

Find $\dfrac{dy}{dx}$ for each of the following: 1) $y = \dfrac{x^2 + 1}{e^x + x^2}$ 2) $y = \left(\dfrac{x^4 + 1}{x^4 - 1}\right)^2$

Answers 1) $\dfrac{2xe^x - x^2 e^x - e^x - 2x}{(e^x + x^2)^2}$ 2) $\dfrac{-16x^3(x^4 + 1)}{(x^4 - 1)^3}$

Derivatives: Implicit Differentiation

Don't panic! Implicit Differentiation is just an application of the **chain rule** in disguise.

Example 1) Let $xy + \sin y = 4$. (a) Find $\dfrac{dy}{dx}$.   (b) Find y and $\dfrac{dy}{dx}$ when $y = \pi$.

Solution It is **impossible** to rewrite this equation with y on the left side and only terms involving x on the right side. (Try to isolate y. Futile!) Instead, to find $\dfrac{dy}{dx}$, we take the derivative with respect to x directly from the equation using the **cardinal rule of math equations** :

| **What you do to one side you must do to the other!** |

Note that we need the **product rule** to deal with the xy term.

$$\frac{d(Left\ Side)}{dx} = \frac{d(Right\ Side)}{dx} \quad \therefore \quad x\frac{dy}{dx} + y(1) + \cos y \frac{dy}{dx} = 0.$$

Now we factor out the $\dfrac{dy}{dx}$ terms and solve: $\dfrac{dy}{dx}(x + \cos y) = -y$ and so $\dfrac{dy}{dx} = \dfrac{-y}{x + \cos y}$

(b) $y = \pi \Rightarrow \pi x + 0 = 4 \Rightarrow x = \dfrac{4}{\pi}$. At the point $\left(\dfrac{4}{\pi}, \pi\right)$, $\dfrac{dy}{dx} = \dfrac{-\pi}{\dfrac{4}{\pi} + \cos(\pi)} = \dfrac{-\pi}{\dfrac{4}{\pi} - 1} = \dfrac{-\pi^2}{4 - \pi}$

Example 2) Find $\dfrac{dy}{dx}$ if $x^2 y^3 + 2x + 3y = \sin(xy) + 4$.

Solution $x^2(3y^2)\dfrac{dy}{dx} + 2xy^3 + 2 + 3\dfrac{dy}{dx} = \cos(xy)\left(x\dfrac{dy}{dx} + y\right)$

| Now bring $\dfrac{dy}{dx}$ terms to the left and factor out $\dfrac{dy}{dx}$. All other terms go to, or stay on, the right side. |

$$\frac{dy}{dx}\left(3x^2 y^2 + 3 - x\cos(xy)\right) = y\cos(xy) - 2 - 2xy^3$$

$$\therefore \quad \frac{dy}{dx} = \frac{y\cos(xy) - 2 - 2xy^3}{3x^2 y^2 + 3 - x\cos(xy)}$$

Two for you.

1)(a) Find $\dfrac{dy}{dx}$ if $e^{x+y}+5x+2y=1$.   (b) Find $\dfrac{dy}{dx}$ at $(0,0)$.

2) Find $\dfrac{dy}{dx}$ if $y\sin x+y=x+\tan y$.

Answers 1)(a) $\dfrac{-(e^{x+y}+5)}{e^{x+y}+2}$ (b) -2 2) $\dfrac{1-y\cos x}{\sin x+1-\sec^2 x}$

Derivatives: Implicit Differentiation Second Derivative

Here the key is one simple fact: $\dfrac{d}{dx}\left(\dfrac{dy}{dx}\right) = \dfrac{d^2y}{dx^2}$, that is, the derivative with respect to x of the first derivative (with respect to x) is the second derivative (with respect to x).
Also, the second derivative may seem to get hairy. It is not implicit differentiation that causes this, **but GRADE 5 fractions!**

Example 1) Find $\dfrac{dy}{dx}$ and $\dfrac{d^2y}{dx^2}$ if $e^y - xy = 1$.

Solution Find the first derivative implicitly.

$$e^y\dfrac{dy}{dx} - x\dfrac{dy}{dx} - y(1) = 0 \Rightarrow \dfrac{dy}{dx}(e^y - x) = y \ \text{ and so } \ \dfrac{dy}{dx} = \dfrac{y}{e^y - x}$$

Now, we use the quotient rule to find $\dfrac{d^2y}{dx^2}$.

BUT REMEMBER : $\dfrac{d}{dx}(e^y) = e^y\dfrac{dy}{dx}$ and $\dfrac{d}{dx}(y) \overset{\boxed{\text{Not to insult you but...}}}{=} \dfrac{dy}{dx}$

$$\dfrac{d^2y}{dx^2} = \dfrac{(e^y - x)\dfrac{dy}{dx} - y(e^y\dfrac{dy}{dx} - 1)}{(e^y - x)^2} \overset{\boxed{\text{Factor out } \frac{dy}{dx} \text{ and expand the numerator.}}}{=} \dfrac{\dfrac{dy}{dx}(e^y - x - ye^y) + y}{(e^y - x)^2}$$

$$\overset{\boxed{\text{We already know } \frac{dy}{dx} = \frac{y}{e^y - x}!}}{=} \dfrac{\left(\dfrac{y}{e^y - x}\right)(e^y - x - ye^y) + y}{(e^y - x)^2} \overset{\boxed{\text{Get a common denominator in the numerator.}}}{=} \dfrac{y(e^y - x - ye^y) + y(e^y - x)}{(e^y - x)} \dfrac{1}{(e^y - x)^2}$$

$$\overset{\boxed{\text{Expand the numerator...}}}{=} \dfrac{ye^y - xy - y^2e^y + ye^y - xy}{(e^y - x)^3} \overset{\boxed{\text{...and collect like terms.}}}{=} \dfrac{2ye^y - 2xy - y^2e^y}{(e^y - x)^3} \ \textbf{Whew!}$$

Alternate Solution Go back to $\dfrac{dy}{dx}(e^y - x) = y$ and differentiate the equation using the Product Rule

$$\dfrac{dy}{dx}(e^y\dfrac{dy}{dx} - 1) + (e^y - x)\dfrac{d^2y}{dx^2} = \dfrac{dy}{dx} \ \text{ and so } \ (e^y - x)\dfrac{d^2y}{dx^2} = 2\dfrac{dy}{dx} - e^y\left(\dfrac{dy}{dx}\right)^2 = 2\dfrac{y}{e^y - x} - e^y\left(\dfrac{y}{e^y - x}\right)^2$$

$$\overset{\boxed{\text{Common Denominator!}}}{=} \dfrac{2ye^y - 2xy - ye^y}{(e^y - x)^2} \ \text{ and so (\textbf{whew} once more), } \ \dfrac{d^2y}{dx^2} = \dfrac{2ye^y - 2xy - y^2e^y}{(e^y - x)^3}$$

Two for you.

Find $\dfrac{d^2y}{dx^2}$ in each of the following: 1) $x^3 + y^3 = 1$ 2) $y^2 + 2xy = 10$

(Hint: in each question, at the **last step**, use the original equation!)

Answers 1) $\dfrac{dy}{dx} = -\dfrac{x^2}{y^2}$, $\dfrac{d^2y}{dx^2} = \dfrac{-2x}{y^5}$ 2) $\dfrac{dy}{dx} = -\dfrac{y}{x+y}$, $\dfrac{d^2y}{dx^2} = \dfrac{10}{(x+y)^3}$

Easy Integrals/Anti-derivatives

Okay, this page is the most basic of the basic--the anti-derivative/integral formulas you learn in your first calculus semester applied with **NO TRICKS, NO TWISTS, NO COMPLICATIONS**. The diabolical stuff, and the means to deal with it, comes later!

$$\int 0\,dx = C \qquad \int 1\,dx = x + C \qquad \int m\,dx = mx + C$$

$$\int f(x) \pm g(x)\,dx = \int f(x)\,dx \pm \int g(x)\,dx \qquad \int cf(x)\,dx = c\int f(x)\,dx$$

$$\int x^n\,dx = \frac{x^{n+1}}{n+1} + C,\ n \neq -1 \qquad \int \frac{1}{x}\,dx = \int x^{-1}\,dx = \ln|x| + C$$

$$\int \sin x\,dx = -\cos x + C \qquad \int \cos x\,dx = \sin x + C$$

$$\int \tan x\,dx = -\ln|\cos x| + C \qquad \int \cot x\,dx = \ln|\sin x| + C$$

$$\int \csc x\,dx = \ln|\csc x - \cot x| + C \qquad \int \sec x\,dx = \ln|\sec x + \tan x| + C$$

$$\int e^x\,dx = e^x + C \qquad \int a^x\,dx = \frac{a^x}{\ln a} + C$$

Example 1) Evaluate the following integrals:

(a) $\int 1 + x^2 + x^{1/3} + x^{-5/3}\,dx$ (b) $\int \sin x + 2\csc x + \frac{1}{3}e^x + \frac{1}{x} + 10^x\,dx$

Solution

(a) $\int 1 + x^2 + x^{1/3} + x^{-5/3}\,dx \overset{\boxed{\int x^n dx = \frac{x^{n+1}}{n+1} + C,\ n \neq -1}}{=} x + \frac{x^3}{3} + \frac{x^{4/3}}{\frac{4}{3}} + \frac{x^{-2/3}}{-\frac{2}{3}} + C = \frac{x^3}{3} + \frac{3x^{4/3}}{4} - \frac{3x^{-2/3}}{2} + C$

(b) $\int \sin x + 2\csc x + \frac{1}{3}e^x + \frac{1}{x} + 10^x\,dx = -\cos x + 2\ln|\csc x - \cot x| + \frac{1}{3}e^x + \ln|x| + \frac{10^x}{\ln 10} + C$

Example 2) If $\int f(x)\,dx = 3x^2 - 2x + C$ and $\int g(x)\,dx = 2\sin x - e^x + D$, find $\int 3f(x) - 2g(x)\,dx$

Solution $\int 3f(x) - 2g(x)\,dx = 3\int f(x)\,dx - 2\int g(x)\,dx = 3(3x^2 - 2x + C) - 2(2\sin x - e^x + D)$

$= 9x^2 - 6x - 4\sin x + 2e^x + 3C - 2D \overset{\boxed{\text{Replace the constant } 3C-2D \text{ with the single constant } E.}}{=} 9x^2 - 6x - 4\sin x + 2e^x + E$

Two for you.

Evaluate the following integrals:

1) $\int -1 + 4x^9 - x^{-5/3}\, dx$
2) $\int \cot x - \sec x - 2^x \ln 2\, dx$ $\left(\text{Hint: } \int 2^x \ln 2\, dx = \ln 2\left(\int 2^x\, dx\right)\right)$

Answers 1) $-x + \dfrac{2x^{10}}{5} + \dfrac{3x^{-2/3}}{2} + C$
2) $\ln|\sin x| - \ln|\sec x + \tan x| - 2^x + C$

Easy Integrals that Need a Little Tweaking

Here is a variety of really easy integrals--except you have to do something first to see why they are so easy.

Example 1) Evaluate the following integrals:

(a) $\int (x^3 + 3)^2 \, dx$

(b) $\int (x^{1/2} + 1)(2x + 5) \, dx$

(c) $\int \frac{x^4 - 7}{x^2} \, dx$

(d) $\int \frac{1}{\sec x} \, dx$

(e) $\int \frac{1}{e^x} \, dx$

Solution (a) **EXPAND**: $\int (x^3 + 3)^2 \, dx = \int x^6 + 6x^3 + 9 \, dx = \frac{x^7}{7} + \frac{3x^4}{2} + 9x + C$

Note how unpleasant this example would be if changed to $\int (x^3 + 3)^{20} \, dx$. Note how much MORE unpleasant it would be if changed to $\int (x^3 + 3)^{2/3} \, dx$! The first you could expand, but would you want to? The second, you can't expand, at least not with your current math toolkit!

But test your understanding of the Chain Rule in Reverse (CRIR):

both $\int \boxed{x^2}(x^3 + 3)^{20} \, dx$ and $\int \boxed{x^2}(x^3 + 3)^{2/3} \, dx$ are **easy**!

$$\int x^2 (x^3 + 3)^{20} \, dx \overset{\boxed{\text{Adjust by 3.}}}{=} \frac{1}{3} \int 3x^2 (x^3 + 3)^{20} \, dx \overset{\boxed{\text{CRIR!}}}{=} \frac{(x^3 + 3)^{21}}{63} + C$$

$$\int x^2 (x^3 + 3)^{2/3} \, dx \overset{\boxed{\text{Adjust by 3.}}}{=} \frac{1}{3} \int 3x^2 (x^3 + 3)^{2/3} \, dx \overset{\boxed{\text{CRIR!}}}{=} \frac{(x^3 + 3)^{5/3}}{5} + C$$

(b) $\int (x^{1/2} + 1)(2x + 5) \, dx \overset{\boxed{\text{Expand!}}}{=} \int 2x^{3/2} + 5x^{1/2} + 2x + 5 \, dx = \frac{4}{5}x^{5/2} + \frac{10}{3}x^{3/2} + x^2 + 5x + C$

(c) $\int \frac{x^4 - 7}{x^2} \, dx \overset{\boxed{\text{Make separate fractions!}}}{=} \int x^2 - 7x^{-2} \, dx = \frac{x^3}{3} - \frac{7x^{-1}}{-1} + C = \frac{x^3}{3} + \frac{7}{x} + C$

(d) $\int \frac{1}{\sec x} \, dx \overset{\boxed{\text{This is just a case of the teacher being sneaky!}}}{=} \int \cos x \, dx = \sin x + C$

(e) $\int \frac{1}{e^x} \, dx \overset{\boxed{\text{This is just a case of the teacher being sneaky again!}}}{=} \int e^{-x} \, dx \overset{\boxed{\text{CRIR: Adjust by } -1.}}{=} - \int -e^{-x} \, dx = -e^{-x} + C$

Two for you.

Evaluate the following integrals:

1) $\int (2x^2 - 1)^3 \, dx$ 2) $\int \dfrac{e^{3x} - 2e^x + e^{-x}}{2e^x} \, dx$

Answers 1) $\dfrac{8x^7}{7} - \dfrac{12x^5}{5} + 2x^3 - x + C$ **2)** $\dfrac{1}{4}e^{2x} - x - \dfrac{1}{4e^{2x}} + C$

The Chain Rule in Reverse: No Adjustments Needed!

Many students find this **very** hard but in fact it is pretty easy. **The chain rule in reverse:**

BASIC $\int f'(x)\,dx = f(x)+C$ **CHAIN RULE IN REVERSE** $\int f'(u)\dfrac{du}{dx}\,dx = f(u)+C$

It's gross but I tell my students that when taking the derivative of $y = f(u)$
–please read this as "f **AT** u"; otherwise, we go to a whole other level of gross! –
with respect to x, the chain rule "**spits out**" du/dx. So, when taking the anti-derivative
or integral with respect to x, the du/dx is "**sucked**" back inside the u!

Think **"DOUBLE S"**: **Spit** for the derivative, **Suck** for the integral. This is not to say that
you are allowed to spit on your derivatives, nor that integrals s…well, let's not finish that
thought. Here are all the basic formulas, modified to show the CRIR.

$$\int u^n \frac{du}{dx}\,dx = \frac{u^{n+1}}{n+1}+C,\, n \neq -1 \qquad \int \frac{1}{u}\frac{du}{dx}\,dx = \int u^{-1}\frac{du}{dx}\,dx = \ln|u|+C$$

$$\int \sin u \frac{du}{dx}\,dx = -\cos u + C \qquad \int \cos u \frac{du}{dx}\,dx = \sin u + C \qquad \int \tan u \frac{du}{dx}\,dx = -\ln|\cos u|+C$$

$$\int \csc u \frac{du}{dx}\,dx = \ln|\csc u - \cot u|+C \qquad \int \sec x \frac{du}{dx}\,dx = \ln|\sec u + \tan u|+C$$

$$\int \cot u \frac{du}{dx}\,dx = \ln|\sin u|+C \qquad \int e^u \frac{du}{dx}\,dx = e^u + C \qquad \int a^u \frac{du}{dx}\,dx = \frac{a^u}{\ln a}+C$$

Example 1) Evaluate these integrals, using the chain rule in reverse. In each, identify u and $\dfrac{du}{dx}$.

(a) $\int (x^3+1)^{10}\,(3x^2)\,dx$ (b) $\int e^{\tan x}\sec^2 x\,dx$ (c) $\int \cos(x^{1/2})\left(\dfrac{1}{2}x^{-1/2}\right)dx$ (d) $\int \dfrac{2x}{x^2+1}\,dx$

Solution (a) $\int (x^3+1)^{10}\,(3x^2)\,dx$ $\overset{\boxed{u=x^3+1,\,\frac{du}{dx}=3x^2}}{=} \dfrac{(x^3+1)^{11}}{11}+C$

(b) $\int e^{\tan x}\sec^2 x\,dx$ $\overset{\boxed{u=\tan x,\,\frac{du}{dx}=\sec^2 x}}{=} e^{\tan x}+C$ (c) $\int \cos(x^{1/2})\left(\dfrac{1}{2}x^{-1/2}\right)dx$ $\overset{\boxed{u=x^{1/2},\,\frac{du}{dx}=\frac{1}{2}x^{-1/2}}}{=} \sin(x^{1/2})+C$

(d) $\int \dfrac{2x}{x^2+1}\,dx$ $\overset{\boxed{u=x^2+1,\,\frac{du}{dx}=2x}}{=} \ln(x^2+1)+C$ (Note: $x^2+1 > 0$, so we don't need absolute value.)

Two for you.

Evaluate these integrals using the **Chain Rule In Reverse**. For each, identify u and du/dx.

1) $\int (x^5 + 2x^2 + 1)^{2/3}(5x^4 + 4x)\,dx$

2) $\int \sin(\sin x)(\cos x)\,dx$

Answers 1) $\dfrac{3}{5}(x^5 + 2x^2 + 1)^{5/3} + C$; $\quad u = x^5 + 2x^2 + 1, \quad du/dx = 5x^4 + 4x$

2) $-\cos(\sin x) + C$; $\quad u = \sin x, \quad du/dx = \cos x$

The Chain Rule in Reverse: Adjustments Needed
BUT Don't Use Substitution!

Now let's use the **the chain rule in reverse** where we adjust by a "**multiplicative constant**". You can always pull a multiplicative constant outside the integral and you can adjust an integral by a multiplicative constant **providing you compensate**. The CRIR "sucks" in the du/dx. The key here is that you recognize **in advance** that you have both the u and the du/dx – at least up to the constant – in the integral. You know before you start that the CRIR will solve the problem.

Example 1) Evaluate these integrals, using the chain rule in reverse. In each, identify u and $\dfrac{du}{dx}$

(a) $\displaystyle\int x^2 (x^3+1)^{10}\ dx$ (b) $\displaystyle\int \frac{\cos\sqrt{x}}{\sqrt{x}}\,dx$ (c) $\displaystyle\int 3e^{\tan x}\sec^2 x\,dx$ (d) $\displaystyle\int \frac{x}{x^2+1}\,dx$

Solution

(a) $\displaystyle\int x^2(x^3+1)^{10}\ dx$ Here, $u = x^3+1$ and $\dfrac{du}{dx} = 3x^2$. We need to multiply by 3. To compensate,

we divide by 3 outside the integral sign.

YOU CAN ALWAYS ADJUST BY A MULTIPLICATIVE CONSTANT.

$$\int x^2(x^3+1)^{10}\ dx = \frac{1}{3}\int 3x^2(x^3+1)^{10}\ dx = \frac{1}{3}\frac{(x^3+1)^{11}}{11} + C = \frac{1}{33}(x^3+1)^{11} + C$$

> (Question: Where has the $3x^2$ gone? Answer: **SUCKED INSIDE by the chain rule!**)

(b) $\displaystyle\int\frac{\cos\sqrt{x}}{\sqrt{x}}\,dx$ Here, $u = \sqrt{x} = x^{1/2}$ and $\dfrac{du}{dx} = \dfrac{1}{2}x^{-1/2} = \dfrac{1}{2\sqrt{x}}$. We need to divide by 2.

To compensate, we multiply by 2 outside the integral sign.

$$\int\frac{\cos\sqrt{x}}{\sqrt{x}}\,dx = 2\int\frac{1}{2}x^{-1/2}\cos\left(x^{1/2}\right)dx = 2\sin\left(x^{1/2}\right)+C = 2\sin\sqrt{x}+C$$

(c) $\displaystyle\int 3e^{\tan x}\sec^2 x\,dx$ Here, $u = \tan x$ and $\dfrac{du}{dx} = \sec^2 x$, so no compensation needed.

But move the constant 3 outside: $\displaystyle\int 3e^{\tan x}\sec^2 x\,dx = 3\int e^{\tan x}\sec^2 x\,dx = 3e^{\tan x} + C$

(d) $\displaystyle\int\frac{x}{x^2+1}\,dx$ Here, $u = x^2+1$ and $\dfrac{du}{dx} = 2x$. Compensate with 2.

$$\int\frac{x}{x^2+1}\,dx = \frac{1}{2}\int\frac{2x}{x^2+1}\,dx \overset{\boxed{x^2+1>0 \text{ so we don't need absolute value.}}}{=} \frac{1}{2}\ln(x^2+1)+C$$

Two for you.

Evaluate: 1) $\int (x^4 + 1) \sin(x^5 + 5x)\, dx$ 2) $\int e^{\sec(2x+1)} \sec(2x+1) \tan(2x+1)\, dx$

Answers 1) $-\dfrac{1}{5}\cos(x^5 + 5x) + C$ 2) $\dfrac{1}{2}e^{\sec(2x+1)} + C$

The Chain Rule in Reverse:
Adjustments Needed and Using Substitution

Let's get something straight: I (**that's right, me, the teacher/author talking!**) don't like using substitution for these problems. Why? Because EVERY TIME you adjust the constant to make things work just right as we did on the preceding page, you consolidate further your understanding of the CRIR. It is so easy! Yet students find it so hard! The real key is recognizing that the questions are **"cooked"**! The du/dx term, up to a constant, **must** be present for most problems or the integral is, in many many many examples, too hard or even undo-able. Even so, this page is almost exactly the same as the preceding one, except here we will (**grumble grumble**) use substitution.

Example 1) Evaluate these integrals, using the chain rule in reverse. In each, identify u and du.

(a) $\int x^2 (x^3 + 1)^{10} \, dx$ (b) $\int \dfrac{\cos \sqrt{x}}{\sqrt{x}} \, dx$ (c) $\int 3e^{\tan x} \sec^2 x \, dx$ (d) $\int \dfrac{x}{x^2 + 1} \, dx$

Solution

(a) $\int x^2 (x^3 + 1)^{10} \, dx$ Here, $u = x^3 + 1$ and $\dfrac{du}{dx} = 3x^2$ $\therefore$ $du = 3x^2 dx$. We need to replace $x^2 dx$.

Since $x^2 dx = \dfrac{1}{3} du$, we have $\int x^2 (x^3 + 1)^{10} \, dx$ [Pull 1/3 outside the integral.] $= \dfrac{1}{3} \int u^{10} \, du = \dfrac{1}{3} \dfrac{u^{11}}{11} + C$ [$u = x^3+1$] $= \dfrac{1}{33} (x^3 + 1)^{11} + C$

(b) $\int \dfrac{\cos \sqrt{x}}{\sqrt{x}} \, dx$ Here, $u = \sqrt{x} = x^{1/2}$ and $\dfrac{du}{dx} = \dfrac{1}{2} x^{-1/2} = \dfrac{1}{2\sqrt{x}}$ $\therefore$ $du = \dfrac{dx}{2\sqrt{x}}$ and so $2du = \dfrac{dx}{\sqrt{x}}$

$\int \dfrac{\cos \sqrt{x}}{\sqrt{x}} \, dx = 2 \int \cos u \, du = 2 \sin u + C = 2 \sin \sqrt{x} + C$

(c) $\int 3e^{\tan x} \sec^2 x \, dx$ Here, $u = \tan x$ and (Let's go right to the differential this time!) $du = \sec^2 x \, dx$

$\int 3e^{\tan x} \sec^2 x \, dx = 3 \int e^u \, du = 3e^u + C = 3e^{\tan x} + C$

(d) $\int \dfrac{x}{x^2 + 1} \, dx$ Here, $u = x^2 + 1$ and $du = 2x \, dx$ $\therefore$ $\dfrac{du}{2} = x \, dx$ and

$\int \dfrac{x}{x^2 + 1} \, dx = \dfrac{1}{2} \int \dfrac{du}{u} = \dfrac{1}{2} \ln |u| + C$ [$x^2+1 > 0$ so we don't need absolute value.] $= \dfrac{1}{2} \ln(x^2 + 1) + C$

Three for you.

Evaluate: 1) $\int \dfrac{x^3 - \csc^2 x}{x^4 + 4\cot x}\,dx$ 2) $\int e^{e^x + x}\,dx$ (Hint: $e^{e^x + x} = e^{e^x} e^x$) 3) $\int \dfrac{x}{(5x^2 + 1)^4}\,dx$

Answers 1) $\dfrac{1}{4}\ln|x^4 + 4\cot x| + C$ 2) $e^{e^x} + C$ 3) $-\dfrac{1}{30(5x^2 + 1)^3} + C$

204

Substitution when the CRIR Doesn't Apply

Here is a question just waiting for the Chain Rule in Reverse: $\int x^3(x^4+1)^{1/2}\,dx$.

A simple adjustment of the multiplicative constant gives a final answer in two steps. Substitution not needed!

$$\int x^3(x^4+1)^{1/2}\,dx = \frac{1}{4}\int 4x^3(x^4+1)^{1/2}\,dx = \frac{1}{6}(x^4+1)^{3/2}+C$$

Now here is a question where the CRIR just doesn't apply: $\int x^2(x+3)^{1/2}\,dx$.

We do not have $\dfrac{du}{dx}$ up to a multiplicative constant. And we can't just expand because of the exponent "1/2". So, we make the $x+3$ the variable by substitution.

Example 1) Evaluate $\int x^2(x+3)^{1/2}\,dx$.

Solution Let $u = x+3$. Then $du = dx$ and $x = u-3$. Therefore,

$$\int x^2(x+3)^{1/2}\,dx = \int (u-3)^2 u^{1/2}\,du \overset{\boxed{\text{Now expand.}}}{=} \int (u^2-6u+9)u^{1/2}\,du = \int u^{5/2}-6u^{3/2}+9u^{1/2}\,du$$

$$=\frac{2}{7}u^{7/2}-\frac{12}{5}u^{5/2}+\frac{3}{2}u^{3/2}+C \overset{\boxed{\text{Resubstitute for } u \text{ in terms of } x.}}{=} \frac{2}{7}(x+3)^{7/2}-\frac{12}{5}(x+3)^{5/2}+\frac{3}{2}(x+3)^{3/2}+C$$

Example 2) Evaluate $\int \dfrac{x+3}{2x-1}\,dx$.

Solution Let $u = 2x-1$. Then $du = 2dx$ so $dx = \dfrac{du}{2}$. Also, $2x = u+1$ and so $x = \dfrac{u+1}{2}$. Therefore

$$\int \frac{\frac{u+1}{2}+3}{u}\,\frac{du}{2} \overset{\boxed{\text{Get a common denominator in the top.}}}{=} \int \frac{\frac{u+1+6}{2}}{2u}\,du \overset{\boxed{\substack{\text{Simplify the fraction.} \\ \text{Remember: } \frac{\left(\frac{a}{b}\right)}{c}=\frac{a}{b}\cdot\frac{1}{c}=\frac{a}{bc}}}}{=} \int \frac{u+7}{4u}\,du \overset{\boxed{\substack{\text{I like to pull the constant} \\ \text{outside the integral.}}}}{=} \frac{1}{4}\int \frac{u+7}{u}\,du$$

$$\overset{\boxed{\text{Make separate fractions.}}}{=} \frac{1}{4}\int 1+\frac{7}{u}\,du = \frac{1}{4}(u+7\ln|u|)+C \overset{\boxed{\text{Substitute } u=2x-1.}}{=} \frac{1}{4}(2x-1+7\ln|2x-1|)+C$$

$$\overset{\boxed{\text{optional}}}{=} \frac{x}{2}-\frac{1}{4}+\frac{7}{4}\ln|2x-1|+C \overset{\boxed{\text{optional: } D=C-\frac{1}{4}!}}{=} \frac{x}{2}+\frac{7}{4}\ln|2x-1|+D$$

205

Two for you.

Evaluate the following integrals: 1) $\int (3x-1)(x+5)^{1/3}\,dx$ 2) $\int \dfrac{2x+x^2}{x+1}\,dx$

Answers 1) $\dfrac{9}{7}(x+5)^{7/3} - 12(x+5)^{4/3} + C$ 2) $\dfrac{x^2}{2} + x - \ln|x+1| + C$

The Derivative of an Integral

This is an application of my favourite Theorem: **The Fundamental Theorem of Calculus--** "the" Fundamental Theorem. **The big one!** In this application, we take the derivative of an integral. You might expect the answer to be the original question. It is. Almost.

Example 1) Evaluate each of the following: (a) $\dfrac{d}{dx}\left(\displaystyle\int_{t=1}^{x} t^3 + 1\ dt\right)$ (b) $\dfrac{d}{dx}\left(\displaystyle\int_{t=1}^{x} t^3 + t + 5\ dt\right)$

Solution (a) $\dfrac{d}{dx}\left(\displaystyle\int_{t=1}^{x} t^3 + 1\ dt\right) = \dfrac{d}{dx}\left(\dfrac{t^4}{4} + t\right)\Big|_{1}^{x} = \dfrac{d}{dx}\left(\dfrac{x^4}{4} + x - \left(\dfrac{1}{4} + 1\right)\right) = x^3 + 1$

So we are back where we started except that the t is replaced by x! Knowing this is what happens...

(b) $\dfrac{d}{dx}\left(\displaystyle\int_{t=1}^{x} t^3 + t + 5\ dt\right)\overset{\boxed{\text{The answer is the original integrand with } t \text{ replaced by } x!}}{=} x^3 + x + 5$

Example 2) (a) $\dfrac{d}{dx}\left(\displaystyle\int_{t=\sin x}^{\tan x} t^3 + 1\ dt\right)$ (b) $\dfrac{d}{dx}\left(\displaystyle\int_{t=B(x)}^{T(x)} f(t)\ dt\right)$ (T for *Top*, B for *Bottom*)

Solution (a) $\dfrac{d}{dx}\left(\displaystyle\int_{t=\sin x}^{\tan x} t^3 + 1\ dt\right)\overset{\boxed{\text{This is just like Example 1(a) so far.}}}{=} \dfrac{d}{dx}\left(\dfrac{t^4}{4} + t\right)\Big|_{\sin x}^{\tan x} = \dfrac{d}{dx}\left(\dfrac{\tan^4 x}{4} + \tan x - \left(\dfrac{\sin^4 x}{4} + \sin x\right)\right)$

$\boxed{\text{NOW WE NEED THE CHAIN RULE!}}$

$= \tan^3 x\ (\sec^2 x) + \sec^2 x - \left(\sin^3 x(\cos x) + \cos(x)\right)$

$\boxed{\text{Factor out the "}\sec^2 x\text{" and "}\cos(x)\text{" terms.}}$

$= (\tan^3 x + 1)\sec^2 x - \left(\sin^3 x + 1\right)\cos(x)$

This time, we come back to the original integrand, but "t" is replaced **first** with "$\tan x$" and the chain rule made us multiply by $\dfrac{d(\tan x)}{dx} = \sec^2 x$, then by "$\sin x$" and the chain rule made us multiply by $\dfrac{d(\sin x)}{dx} = \cos x$. So, knowing that this is what happens...

(b) $\dfrac{d}{dx}\left(\displaystyle\int_{t=B(x)}^{T(x)} f(t)\ dt\right) = f(T(x))\dfrac{d(T(x))}{dx} - f(B(x))\dfrac{d(B(x))}{dx}\overset{\boxed{\text{in short...}}}{=} f(T)T' - f(B)B'$

Two for you.

Evaluate the following: (a) $\dfrac{d}{dx}\left(\displaystyle\int_{t=e^x}^{x^2+1} \cos t \ \ln t + t \ dt \right)$ (b) $\dfrac{d}{dx}\left(\displaystyle\int_{\ln x}^{x} x + \ln x \ dx \right)$

Answers 1) $2x\left(\cos(x^2+1) \ \ln(x^2+1) + x^2 + 1 \right) - e^x\left(x\cos\left(e^x\right) + e^x \right)$ (Note: remember $\ln(e^x) = x$.)

2) $x + \ln x - \dfrac{\ln x + \ln(\ln x)}{x}$ (Note: the variable in the integrand, "x" or "t", **does not matter!**)

Finding the Inverse of a Function

Let $f(x) = x^3 + 1$. Then $f(2) = 9$. In the inverse function, we should have $f^{-1}(9) = 2$. In other words, since the point $(2, 9)$ satisfies f, the point $(9, 2)$ must satisfy f^{-1}. The key to inverses is that **the roles of x and y are INTERCHANGED!** So to find the inverse of a function, we 1) **interchange x and y** and then 2) **solve for y.**

Example 1)(a) Find the inverse function for $f(x) = x^3 + 1$.

(b) Verify that $f^{-1}(f(x)) = x$ for $x \in$ domain(f) and $f(f^{-1}(x)) = x$ for $x \in$ domain(f^{-1}).

(c) In general, how are the domains and ranges of f and f^{-1} related?

(d) Draw the graphs of f, f^{-1}, and $y = x$ on the same set of axes. How are the three graphs related?

Solution (a) First, write $y = x^3 + 1$. Interchange x and y:

$x = y^3 + 1$. Now, solve for y: $y^3 = x - 1$ and so

$y = f^{-1}(x) = (x-1)^{1/3}$. Note $f^{-1}(9) = (9-1)^{1/3} = 8^{1/3} = 2$. It works!

(b) $x \in$ dom(f): $f^{-1}(f(x)) = f^{-1}(x^3 + 1) = \left((x^3 + 1) - 1\right)^{1/3} = \left(x^3\right)^{1/3} = x$

$x \in$ dom(f^{-1}): $f(f^{-1}(x)) = f((x-1)^{1/3}) = \left((x-1)^{1/3}\right)^3 + 1 = x - 1 + 1 = x$

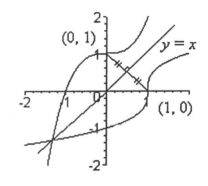

(c) domain(f) = range(f^{-1})   range(f) = domain(f^{-1})

(d) f and f^{-1} are mirror images in the line $y = x$!

For example, look at points $(1, 0)$ and $(0, 1)$.

Note : **Given the function $y = f(x)$, the inverse will be a function only if f is one to one!**
For example, if $f(x) = x^2$, then the inverse **relation** is $f^{-1}(x) = \pm\sqrt{x}$. Since $f(3) = f(-3) = 9$, therefore $f^{-1}(9) = \pm 3$; f^{-1} is **not** a function!

Example 2)(a) Find the inverse of $y = f(x) = \ln(x + 1)$.

(b) State the domain and range for each of f and f^{-1}.

Solution (a) Set $x = \ln(y + 1)$. $\therefore$ $y + 1 = e^x$ and so $y = f^{-1}(x) = e^x - 1$.

(b) Domain of f: $x + 1 > 0$ and so $x > -1$. Range of f: $y \in R$.

Domain of f^{-1}: $x \in R$. Range of f^{-1}: $y = e^x - 1 > -1$.

The domains and ranges of f and f^{-1} are interchanged!

Two for you.

Find f^{-1} for each of the following: 1) $f(x) = x^5 + 5$ 2) $f(x) = e^{2x+}$

Answers 1) $f^{-1}(x) = (x-5)^{1/5}$ 2) $f^{-1}(x) = \dfrac{\ln x - 1}{2}$

Derivatives of Inverse Functions

Let $y = f(x)$. Then $\dfrac{dy}{dx} = f'(x)$. For the inverse function, we want x and y to switch roles,

so the derivative of the inverse should be, at least in notation, $\dfrac{dx}{dy}$. But remember that we can

treat dx and dy as separate quantities (run and rise along the tangent line – see **Estimating
Using the Differential** in your Survival Kit!). Therefore, from Grade 5 arithmetic,

$$\dfrac{dx}{dy} \overset{\text{should}}{=} \dfrac{1}{\left(\dfrac{dy}{dx}\right)}.$$ **AND IT DOES!** But there is a **subtle** part: remember we have interchanged x and y

We evaluate $\dfrac{dx}{dy}$ at the point (y, x) while we evaluate $\dfrac{dy}{dx}$ at the point (x, y).

The derivative of the inverse function at the point (y, x) is the reciprocal of the derivative of the original function at the point (x, y).

Example 1) Let $y = f(x) = x^3 + 1$. (a) Find $f^{-1}(x)$, $f'(x)$, and $(f^{-1})'(x)$.

(b) Note that $f(2) = 9$. Verify that $(f^{-1})'(9) = \dfrac{1}{f'(2)}$.

Solution (a) In $y = f(x) = x^3 + 1$, interchange x and y: $x = y^3 + 1$. Now solving for y,

$y^3 = x - 1$ and so $y = f^{-1}(x) = (x-1)^{1/3}$. Therefore, $f'(x) = 3x^2$ and $(f^{-1})'(x) = \dfrac{1}{3(x-1)^{2/3}}$.

(b) $f'(2) = 12$ and $(f^{-1})'(9) = \dfrac{1}{3(8)^{2/3}} = \dfrac{1}{12} = \dfrac{1}{f'(2)}$

Two for you.

Find $(f^{-1})'$ at each of the points 1) $(2, 8)$ and 2) $(-2, 8)$ for the function $y = f(x) = x^2 + 4$.

Answers 1) $\dfrac{1}{4}$　　2) $-\dfrac{1}{4}$

(Note the problem with asking for "$(f^{-1})'(8)$"! Since $f(x)$ is not a one to one function, f^{-1} is **NOT A FUNCTION!** For example, $f(2) = f(-2) = 8$.

We **can't** have $(f^{-1})'(8) = \dfrac{1}{f'(2)}$ **and** $(f^{-1})'(8) = \dfrac{1}{f'(-2)}$.)

Polar Coordinates

POSITIVE angles are drawn **COUNTER-CLOCKWISE** from the positive x axis.

NEGATIVE angles are drawn **CLOCKWISE** from the positive x axis.

POSITIVE RADIUS : Given the point P with polar coordinates (r, θ), with $r > 0$, plot P, with the angle θ, r units from the pole (or origin).

NEGATIVE RADIUS : Given the point P with polar coordinates (r, θ), with $r < 0$, plot a point **(not P!)**, with the angle θ, $-r$ units from the pole (or origin). Then **reflect** this point through the pole. **The reflected point is P!**

Example 1) Plot each of the following points which are given in polar coordinates.

(a) $(3, \pi/4)$ (b) $(1, -\pi/3)$ (c) $(2, 0)$ (d) $(2, \pi)$

Solution

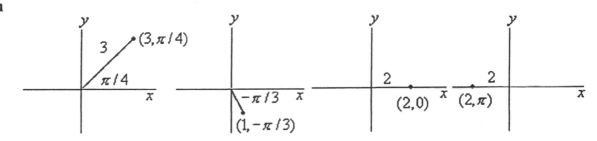

Example 2) Plot each of the following points which are given in polar coordinates.

(a) $(-3, \pi/4)$ (b) $(-1, -\pi/3)$ (c) $(-2, 0)$ (d) $(-2, \pi)$

Solution

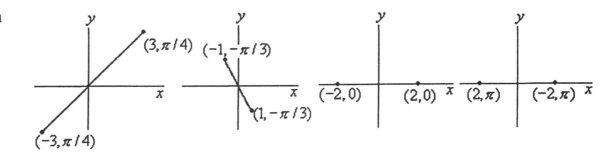

213

Two for you.

1) Plot the points with polar coordinates $(2, 30°)$ and $(2, -30°)$.

2) Plot the points with polar coordinates $(-2, 30°)$ and $(-2, -30°)$.

Answers

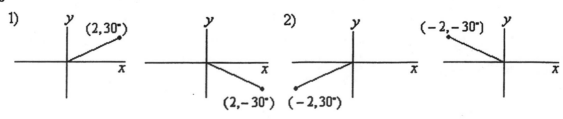

214

Polar to Rectangular Coordinates
Rectangular to Polar Equations

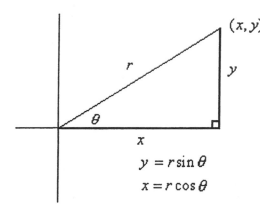

$y = r\sin\theta$

$x = r\cos\theta$

The problem with math teachers when it comes to things like polar and rectangular coordinates is this: how are students to know whether (a, b) is supposed to be in **polar** (a is the radius and b is the angle) coordinates or **rectangular** (a is the horizontal distance and b is the vertical distance) coordinates?

Good answer: teachers would label points something like this: $(a, b)_{PC}$ and $(a, b)_{RC}$!
"Get Real" answer: you are either told explicitly or you pick it up from context.
Sometimes life is cruel and teachers can be (though this is rare) inconsiderate!

Example 1) Convert each of the following points from polar to rectangular coordinates:

(a) $(3, \pi/4)$ (b) $(2, -\pi/3)$ (c) $(-2, 0)$ (d) $(2, \pi/2)$ (e) $(2, -\pi/2)$

Solution (a) $x = 3\cos\pi/4 = 3\left(\dfrac{1}{\sqrt{2}}\right) = \dfrac{3}{\sqrt{2}}$ and $y = 3\sin\pi/4 = 3\left(\dfrac{1}{\sqrt{2}}\right) = \dfrac{3}{\sqrt{2}}$

(b) $x = 2\cos(-\pi/3) = 2\left(\dfrac{1}{2}\right) = 1$ and $y = 2\sin(-\pi/2) = 2\left(-\dfrac{\sqrt{3}}{2}\right) = -\sqrt{3}$

(c) $x = -2\cos 0 = 2(1) = -2$ and $y = -2\sin 0 = 0$
(d) $x = 2\cos(\pi/2) = 0$ and $y = 2\sin(\pi/2) = 2$
(e) $x = 2\cos(-\pi/2) = 0$ and $y = 2\sin(-\pi/2) = -2$

Example 2) Convert these rectangular equations to polar: (a) $y = x^2$ (b) $x^2 + y^2 = 16$

Solution (a) $r\sin\theta = r^2\cos^2\theta$ $\therefore$ $r = 0$ or $r = \dfrac{\sin\theta}{\cos^2\theta} = \tan\theta\sec\theta$.

Note that when $\theta = 0$ in the second equation, $r = 0$. Therefore, we only need $r = \tan\theta\sec\theta$

(b) $r^2\cos^2\theta + r^2\sin^2\theta = 16$ and factoring out r^2, we have $r^2(\cos^2\theta + \sin^2\theta) = 16$.

Everybody knows that $\cos^2\theta + \sin^2\theta = 1$ and so the polar equation is $r^2 = 16$.

Two for you.

1) Convert these polar coordinates to rectangular: (a) $(5, 5\pi/4)$ (b) $(-2, \pi)$

2) Find a polar equation from the rectangular equation $y = x$.

Answers 1)(a) $\left(-\dfrac{5}{\sqrt{2}}, -\dfrac{5}{\sqrt{2}}\right)$ (b) $(2, 0)$ 2) $\tan\theta = 1$

Rectangular to Polar Coordinates
Polar to Rectangular Equations

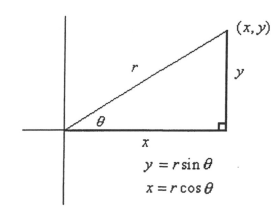

$$y = r\sin\theta$$
$$x = r\cos\theta$$

$r^2 = x^2 + y^2$ and so $r = \pm\sqrt{x^2 + y^2}$ $\quad \tan\theta = \dfrac{y}{x}$

Use your calculator or math software to solve $\tan\theta = $ constant, if it isn't an "easy" ratio.

> **Set your calculator to radians for now :**
> **constant 2nd function tan =**

Be careful! Your calculator will only give you an answer for θ between $-\pi/2$ and $\pi/2$.

Strategy: (x, y) is in one of the four quadrants

OR
is on the positive y axis ($x = 0$, $y > 0$)

OR
is on the negative y axis ($x = 0$, $y < 0$).
The picture at the right shows how to choose r and θ in each case.

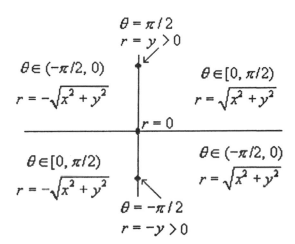

Example 1) Convert each of the following points from rectangular to polar coordinates:

(a) $(2, 2)$ (b) $(-1, \sqrt{3})$ (c) $(0, -5)$

Solution (a) $\tan\theta = \dfrac{2}{2} = 1$ and the point is in the first quadrant. $\therefore$ $r = +\sqrt{2^2 + 2^2} = \sqrt{8} = 2\sqrt{2}$

(b) $\tan\theta = -\sqrt{3}$ and the point is in the second quadrant. $\therefore$ $r = -\sqrt{(-1)^2 + \sqrt{3}^2} = -\sqrt{4} = -2$

(c) $\tan\theta$ is undefined. Since $y < 0$, we can choose $\theta = -\pi/2$ and $r = 5$.

Example 2) Convert $r\sin\theta = 5$ to an equation in rectangular coordinates.

Solution $\pm\sqrt{x^2 + y^2} \left(\dfrac{y}{\pm\sqrt{x^2 + y^2}} \right) = 5$ which becomes simply $y = 5$!

Two for you.

1) Convert rectangular coordinates $(-3, -3)$ to polar coordinates.

2) Find a rectangular equation corresponding to $r = \sin\theta$.

Answers 1) $\theta = \pi/4$ and $r = -3\sqrt{2}$ 2) $x^2 + y^2 = y$ or $x^2 + \left(y - \frac{1}{2}\right)^2 = \frac{1}{4}$

The Dot or Scalar or Inner Product of Two Vectors

Unbelievable. We almost got through the **entire** Survival Kit without mentioning **vectors** once! Well, here is the **dot product** to the rescue. The dot product is also called the **"scalar product"** because it is a form of vector multiplication which yields a scalar, **not a vector!** Why is it called the inner product as well? See ****** below.

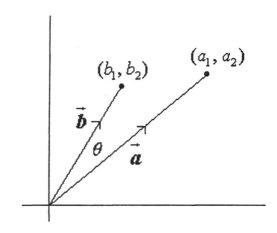

Definition The dot or scalar product of the two vectors $\vec{a}$ and $\vec{b}$ is $\vec{a} \cdot \vec{b} = \|\vec{a}\| \|\vec{b}\| \cos\theta$, where θ is the angle between $\vec{a}$ and $\vec{b}$, $0 \le \theta \le \pi$.

If, as in the diagram, $\vec{a} = (a_1, a_2)$ and $\vec{b} = (b_1, b_2)$, it turns out that

$$\vec{a} \cdot \vec{b} = \|\vec{a}\| \|\vec{b}\| \cos\theta = a_1 b_1 + a_2 b_2.$$

If we are dealing with more dimensions,

$\vec{a} = (a_1, a_2, ..., a_n)$ and $\vec{b} = (b_1, b_2, ..., b_n)$,

then $\vec{a} \cdot \vec{b} = a_1 b_1 + a_2 b_2 + ... + a_n b_n.$**

Example 1) Given $\vec{a} = (3, 4, 0)$ and $\vec{b} = (-1, 2, \sqrt{11})$, find

(a) $\vec{a} \cdot \vec{b}$ (b) $\|\vec{a}\|$ and $\|\vec{b}\|$ (c) θ, the angle between $\vec{a}$ and $\vec{b}$.

Solution (a) $\vec{a} \cdot \vec{b} = (3, 4, 0) \cdot (-1, 2, \sqrt{11}) = 3 \times (-1) + 4 \times 2 + 0 \times \sqrt{11} = 5$

(b) $\|\vec{a}\| = \sqrt{3^2 + 4^2 + 0^2} = \sqrt{25} = 5$ $\|\vec{b}\| = \sqrt{(-1)^2 + 2^2 + \sqrt{11}^2} = \sqrt{16} = 4$

(c) From $\vec{a} \cdot \vec{b} = \|\vec{a}\| \|\vec{b}\| \cos\theta = a_1 b_1 + a_2 b_2 + a_3 b_3$, $\cos\theta = \dfrac{a_1 b_1 + a_2 b_2 + a_3 b_3}{\|\vec{a}\| \|\vec{b}\|} = \dfrac{5}{5 \times 4} = \dfrac{1}{4} = 0.25$

Put your calculator in RADIAN mode!
0.25 SecondFunction Cos =

Put your calculator in DEGREE mode!
0.25 SecondFunction Cos =

$\therefore \theta \doteq 1.32$ radians   or $\therefore \theta \doteq 75.5°$

Example 2) What can you conclude about θ or $\vec{a}$ or $\vec{b}$ if $\vec{a} \cdot \vec{b}$ is (a) 0?   (b) $\|\vec{a}\| \|\vec{b}\|$?

Solution (a) If $\vec{a} \cdot \vec{b} = \|\vec{a}\| \|\vec{b}\| \cos\theta = 0$, then either $\vec{a} = \vec{0}$ or $\vec{b} = \vec{0}$ or $\theta = \pi/2$.

(b) If $\vec{a} \cdot \vec{b} = \|\vec{a}\| \|\vec{b}\| \cos\theta = \|\vec{a}\| \|\vec{b}\|$, then $\cos\theta = 1$ and so $\theta = 0$.

Two for you.

1) Find the angle between $\vec{a}$ and $\vec{b}$ if $\vec{a} = (1, -1)$ and $\vec{b} = (-1, 2)$.

2) What can you conclude about θ or $\vec{a}$ or $\vec{b}$ if the value of $\vec{a} \cdot \vec{b}$ is $-\|\vec{a}\| \|\vec{b}\|$?

Answers 1) $\theta \doteq 2.8$ radians or $\theta \doteq 161.6°$.   2) $\theta = \pi$

The Vector or Cross Product of Two Vectors

The cross product is also called the "**vector product**" because it is a type of vector multiplication which, unlike the dot product, **does yield a vector!** In fact, the cross product of $\vec{a}$ and $\vec{b}$ (which are vectors in $\mathbb{R}^3$) is a vector **perpendicular to the plane (when $\vec{a}$ and $\vec{b}$ are linearly independent) spanned by $\vec{a}$ and $\vec{b}$!**

Definition The cross or vector product of the two vectors $\vec{a}$ and $\vec{b}$ is a vector with magnitude $\|\vec{a} \times \vec{b}\| = \|\vec{a}\| \|\vec{b}\| \sin\theta$, where θ is the angle between $\vec{a}$ and $\vec{b}$, $0 \le \theta \le \pi$. (Note that this guarantees $\|\vec{a} \times \vec{b}\| = \|\vec{a}\| \|\vec{b}\| \sin\theta \ge 0$!) The direction of $\vec{a} \times \vec{b}$ is given by the...

Right Hand Rule : If you let the fingers of your right hand curl in the direction of rotation from $\vec{a}$ to $\vec{b}$, then your thumb points in the direction of $\vec{a} \times \vec{b}$.

$\vec{a} \times \vec{b}$ → Thumb up!

$\vec{a}$ θ $\vec{b}$

Curl the fingers of your right hand.

If $\vec{a} = (a_1, a_2, a_3)$ and $\vec{b} = (b_1, b_2, b_3)$, it turns out that

$$\vec{a} \times \vec{b} = (a_2 b_3 - a_3 b_2,\ a_3 b_1 - a_1 b_3,\ a_1 b_2 - a_2 b_1) \quad \boxed{\text{This is for those of you who are comfortable with the DETERMINANT of a 3×3 matrix.}} = \begin{vmatrix} \vec{i} & \vec{j} & \vec{k} \\ a_1 & a_2 & a_3 \\ b_1 & b_2 & b_3 \end{vmatrix}.$$

> Here is an easy pattern for $\vec{a} \times \vec{b}$! The **1st** coordinate uses 2, 3; the **2nd** uses 3, 1; the **3rd** uses 1, 2.

Example 1) Given $\vec{a} = (1, 2, 3)$ and $\vec{b} = (4, 5, 6)$, find $\vec{a} \times \vec{b}$. Is θ a first or second quadrant angle?

Solution $\vec{a} \times \vec{b} = (1, 2, 3) \times (4, 5, 6) = (2 \times 6 - 5 \times 3,\ 3 \times 4 - 1 \times 6,\ 1 \times 5 - 4 \times 2) = (-3, 6, -3)$

Note that $\vec{a} \cdot \vec{b} = 4 + 10 + 18 = 32 > 0$.

$\therefore \cos\theta \quad \boxed{\text{We used this formula on the previous page!}} = \dfrac{\vec{a} \cdot \vec{b}}{\|\vec{a}\| \|\vec{b}\|} > 0$ and so $\theta \in (0, \pi/2)$.

Example 2) Assume neither $\vec{a} = (a_1, a_2, a_3)$ nor $\vec{b} = (b_1, b_2, b_3)$ is $\vec{0}$ and $\vec{a} \not\parallel \vec{b}$, so $\vec{a} \times \vec{b} \neq \vec{0}$. Show that $\vec{a} \times \vec{b}$ is orthogonal to $\vec{a}$. (Remember that we define $\vec{0}$ to be orthogonal to all vectors.)

Solution $(\vec{a} \times \vec{b}) \cdot \vec{a} = (a_2 b_3 - a_3 b_2,\ a_3 b_1 - a_1 b_3,\ a_1 b_2 - a_2 b_1) \cdot (a_1, a_2, a_3)$

$= a_1 a_2 b_3 - a_1 a_3 b_2 + a_2 a_3 b_1 - a_2 a_1 b_3 + a_3 a_1 b_2 - a_3 a_2 b_1 \quad \boxed{\text{The terms cancel out in pairs!}} = 0$. If θ is the angle between $\vec{a} \times \vec{b}$ and $\vec{a}$, then $(\vec{a} \times \vec{b}) \cdot \vec{a} = \|\vec{a} \times \vec{b}\| \|\vec{a}\| \cos\theta$. $\therefore \cos\theta = 0$ and so $\theta = \pi/2$

Two for you.

1) Find $\vec{a} \times \vec{b}$ if $\vec{a} = (1, 2, 3)$ and $\vec{b} = (1, 2, 3)$.

2) What can you conclude about $\vec{a} \times \vec{b}$, if anything, if $\vec{a}$ is parallel to $\vec{b}$?

Answers 1) $\vec{a} \times \vec{b} = \vec{0}$ 2) $\vec{a} \times \vec{b} = \vec{0}$

Summation Notation and Common $\text{SUM} \equiv \sum$ Formulas

Let m and n be natural numbers with $1 \le m < n$.

$$\sum_{i=1}^{n} f(i) = f(1) + f(2) + f(3) + \ldots + f(n) \qquad \sum_{i=m}^{n} f(i) = f(m) + f(m+1) + f(m+2) + \ldots + f(n)$$

$$\sum_{i=1}^{n} cf(i) = c\sum_{i=1}^{n} f(i) \qquad \sum_{i=1}^{n}(f(i) \pm g(i)) = \sum_{i=1}^{n} f(i) \pm \sum_{i=1}^{n} g(i)$$

$$\sum_{i=1}^{n} c = cn \qquad \sum_{i=1}^{n} i = \frac{n(n+1)}{2} \qquad \sum_{i=1}^{n} i^2 = \frac{n(n+1)(2n+1)}{6} \qquad \sum_{i=1}^{n} i^3 = \frac{n^2(n+1)^2}{4} = \left(\sum_{i=1}^{n} i\right)^2$$

Example 1) Evaluate each of (a) and (b) and expand (c) into THREE sums.

(a) $\displaystyle\sum_{i=1}^{11} 2i$ (b) $\displaystyle\sum_{i=4}^{8} i^3$ (c) $\displaystyle\sum_{i=1}^{n}(1+i)^2$

Solution (a) $\displaystyle\sum_{i=1}^{11} 2i = 2\sum_{i=1}^{11} i = 2\left(\frac{11 \cdot 12}{2}\right) = 132$ (b) $\displaystyle\sum_{i=4}^{8} i^3 \overset{\boxed{\sum_{i=4}^{8} i^3 - \sum_{i=4}^{3} i^3}}{=} \left(\frac{8 \cdot 9}{2}\right)^2 - \left(\frac{3 \cdot 4}{2}\right)^2 = 36^2 - 36 = 1260$

(c) $\displaystyle\sum_{i=1}^{n}(1+i)^2 = \sum_{i=1}^{n}(1 + 2i + i^2) = \sum_{i=1}^{n} 1 + 2\sum_{i=1}^{n} i + \sum_{i=1}^{n} i^2$

Example 2) Write the following in summation notation and evaluate the sum:

$$S = \frac{2}{n}\left(\frac{2}{n}\right)^2 + \frac{2}{n}\left(\frac{4}{n}\right)^2 + \frac{2}{n}\left(\frac{6}{n}\right)^2 + \ldots + \frac{2}{n}\left(\frac{2i}{n}\right)^2 + \ldots + \frac{2}{n}\left(\frac{2n}{n}\right)^2$$

Solution $S = \dfrac{2}{n}\left(\dfrac{2}{n}\right)^2 + \dfrac{2}{n}\left(\dfrac{4}{n}\right)^2 + \dfrac{2}{n}\left(\dfrac{6}{n}\right)^2 + \ldots + \dfrac{2}{n}\left(\dfrac{2i}{n}\right)^2 + \ldots + \dfrac{2}{n}\left(\dfrac{2n}{n}\right)^2$

$$= \sum_{i=1}^{n} \frac{2}{n}\left(\frac{2i}{n}\right)^2 \overset{\boxed{\text{Factor out the } \frac{2}{n} \text{ and the } \left(\frac{2}{n}\right)^2!}}{=} \frac{2}{n}\left(\frac{2}{n}\right)^2 \sum_{i=1}^{n} i^2 = \frac{8}{n^3}\frac{n(n+1)(2n+1)}{6} = \frac{4(n+1)(2n+1)}{3n^2}$$

Note: For evaluating integrals from definition (where a sum like this usually arises), we'd continue...

$$\frac{4(n+1)(2n+1)}{3n^2} = \frac{4}{3}\left(\frac{n+1}{n}\right)\left(\frac{2n+1}{n}\right) = \frac{4}{3}\left(1+\frac{1}{n}\right)\left(2+\frac{1}{n}\right)$$

Two for you.

1) Evaluate $\displaystyle\sum_{i=1}^{10}\left(2i-i^2\right)$.

2) Expand into THREE sums: $\displaystyle\sum_{i=1}^{n}\frac{3}{n}\left(1+\frac{3i}{n}\right)^2$

Answers 1) -275

2) $\displaystyle\frac{3}{n}\sum_{i=1}^{n}1+\frac{18}{n^2}\sum_{i=1}^{n}i+\frac{27}{n^3}\sum_{i=1}^{n}i^2$

Arithmetic and Geometric Sequences and Series

The nth term of an **arithmetic** sequence with first term a_1 and common difference d is

$a_n = a_1 + (n-1)d$. The sum is $S_n = \sum_{i=1}^{n} a_i = \frac{n}{2}(2a_1 + (n-1)d) \overset{\boxed{\text{ALSO}}}{=} \frac{n}{2}(a_1 + a_n)$.

The nth term of a **geometric** sequence with first term a_1 and common ratio r is

$a_n = a_1 r^{n-1}$. The sum is $S_n = \sum_{i=1}^{n} a_i r^{i-1} \overset{\boxed{\text{Use this when } r>1.}}{=} a_1\left(\frac{r^n - 1}{r-1}\right) \overset{\boxed{\text{Use this when } r<1.}}{=} a_1\left(\frac{1-r^n}{1-r}\right)$.

> Here are the corresponding formulas for i starting at 0.
>
> Arithmetic: $a_n = a_0 + nd$, $S_n = \frac{n+1}{2}(a_1 + a_n)$ Geometric: $a_n = a_0 r^n$, $S_n = a_0\left(\frac{1-r^{n+1}}{1-r}\right)$

Example 1) Given an arithmetic series with $a_1 = 7$ and common difference $d = 2$, find a_6 and S_6

Solution $a_6 \overset{\boxed{a_1=7, n=6, d=2}}{=} a_1 + (6-1)2 = 7 + 5(2) = 17$

$S_6 \overset{\boxed{a_1=7, n=6, d=2}}{=} \frac{6}{2}(2(7) + (6-1)2) = 3(14+10) = 72$ **OR** $S_6 = \frac{6}{2}(a_1 + a_6) = 3(7+17) = 72$

Example 2) Given a geometric sequence with first term $a_1 = 3$ and common ratio $r = 2$, find a_6 and S_6.

Solution $a_6 = a_1 r^5 \overset{\boxed{a_1=3, r=2}}{=} 3(2^5) = 96$ $S_6 = a_1\left(\frac{r^6 - 1}{r-1}\right) \overset{\boxed{a_1=3, r=2}}{=} 3\left(\frac{2^6 - 1}{2-1}\right) = 189$

Example 3) An arithmetic sequence has $a_4 = 16$ and $a_{12} = 56$. Find a_1 and d.

Solution $a_{12} = a_1 + 11d = 56$ and $a_4 = a_1 + 3d = 16$. Subtracting these two equations gives $8d = 40$ so $d = 5$. Substituting: $a_4 = a_1 + 15 = 16$ and so $a_1 = 1$.

Example 4) A geometric sequence has $a_4 = 10\ 000$ and $a_7 = 10$. Find a_1 and r.

Solution $a_4 = a_1 r^3 = 10000$ and $a_7 = a_1 r^6 = 10$.

$\dfrac{a_7}{a_4} = \dfrac{a_1 r^6}{a_1 r^3} = r^3 = \dfrac{10}{10\ 000} = \dfrac{1}{1000}$ and so $r = \dfrac{1}{10}$. Substituting in a_4 gives

$a_1\left(\dfrac{1}{10}\right)^3 = 10\ 000$ and so $a_1 = 10\ 000\ 000$.

Two for you.

1) Given an arithmetic sequence with 5th term 50 and 13th term 26, find a_1 and d

2) Given a geometric sequence with 5th term 2 and 12th term 256, find a_1 and r.

.

.

Answers 1) $d = -3$ and $a_1 = 62$ 2) $r = 2$ and $a_1 = \dfrac{1}{8}$

Combinations and Permutations: Choosing and Arranging

For any natural numbers n and r, where $n \geq r$,

$$C(n, r) \overset{\boxed{\text{Another Notation!}}}{=} \binom{n}{r} \overset{\boxed{\text{Another Notation!!}}}{=} {}_nC_r = \frac{n!}{(n-r)!\, r!} = \binom{n}{n-r} \text{ calculates}$$

the number of possible **COMBINATIONS** of r objects you can make from n objects.

$$P(n, r) \overset{\boxed{\text{Another Notation!}}}{=} {}_nP_r = \frac{n!}{(n-r)!} = \text{ calculates}$$

the number of possible **ARRANGEMENTS** of r objects you can make from n objects.

Remember, $0! = 1! = 1$ and $i! = i(i-1)(i-2)...(3)(2)(1)$.

$$C(n, 0) = C(n, n) = 1 \qquad P(n, 0) = 1 \qquad P(n, n) = n!$$

Example 1) How many possible ways are there of choosing a president, vice-president, and treasurer from a group of 8 candidates for a student math club? **(Yes, there are student math clubs!!)**

Solution We are looking for the number of **arrangements** of 3 people from 8. **Order IS important!**

The required number is $P(8,3) = \dfrac{8!}{(8-3)!} = \dfrac{8!}{5!} = 8(7)(6) = 336$.

Example 2) How many possible ways are there of choosing 3 members for the executive of the math club from 8 candidates?

Solution We are looking for the number of **combinations** of 3 people from 8. **Order is NOT important!**

The required number is $C(8,3) = \dfrac{8!}{(8-3)!3!} = \dfrac{8!}{5!3!} = 8(7) = 56$.

Two for you.

1) How many different poker hands (5 cards from a 52 card deck) contain the Ace and Jack of Spades?

2) For the word "utopia", how many possible ways are there of arranging

(a) exactly 4 of all the letters?   (b) just the vowels?

Answers 1) $C(50,3) = 19600$ 2)(a) $P(6,4) = 360$ (b) $P(4,4) = 24$

Mean, Median, Mode, and Standard Deviation

Given numbers $X_1 \le X_2 \le X_3 \le ... \le X_k$, we have **THREE** kinds of averages:

$$\textbf{Mean} = \frac{X_1 + X_2 + X_3 + ... + X_k}{k} = \frac{\sum\limits_{i=1}^{k} X_i}{k}$$

Median The numbers have been listed from lowest to highest. The median is

$$\begin{cases} \text{the middle number } X_{\frac{k+1}{2}}, \text{ if } k \text{ is odd} \\[2ex] \text{the average } \dfrac{X_{\frac{k}{2}} + X_{\frac{k}{2}+1}}{2}, \text{ if } k \text{ is even} \end{cases}$$

Mode The mode is the number that occurs most often in the list.
There may be several modes.

The **standard deviation** is a measure of how the data is scattered about the mean.
If the mean = $\overline{X}$, then

$$\textbf{Standard Deviation} = \sqrt{\frac{\sum\limits_{i=1}^{n}(X_i - \overline{X})^2}{n}}. \text{ Note that } \frac{\sum\limits_{i=1}^{n}(X_i - \overline{X})^2}{n} \text{ is itself a "mean".}$$

It is the **mean of the squares of the distances of the data points to the mean of the original data**
What a mouthful!

Example 1) A group of 13 customers in a women's shoe store have these shoe sizes:
6, 6.5, 7, 7, 7, 7.5, 7.5, 8, 8, 8, 9, 9, 10. Find the mean, median, mode(s), and standard
deviation for this data.

Solution Mean $= \dfrac{6 + 6.5 + 7 \times 3 + 7.5 \times 2 + 8 \times 3 + 9 \times 2 + 10}{13} = \overline{X} \doteq 7.7$

Median $= X_{\frac{13+1}{2}} = X_7 = 7.5$

Mode There are two modes: both 7 and 8 occur three times in the list.

$$\textbf{SD} = \sqrt{\frac{(6 - \overline{X})^2 + (6.5 - \overline{X})^2 + 3(7 - \overline{X})^2 + 2(7.5 - \overline{X})^2 + 3(8 - \overline{X})^2 + 2(9 - \overline{X})^2 + (10 - \overline{X})^2}{13}}$$

$\doteq 1.07$

One for you.

1) Find (a) the mean (b) the median (c) the mode(s) (d) the standard deviation
for the data points 1, 2, 3, 3, 3, 3, 4, 4, 5, 8.

Answers (a) 3.6 (b) $\dfrac{X_5 + X_6}{2} = 3$ (c) 3 (d) 1.8

The Binomial Theorem

For any natural number n,

$$(a+b)^n = \sum_{i=0}^{n}\binom{n}{i}a^{n-i}b^i = \binom{n}{0}a^n + \binom{n}{1}a^{n-1}b + \binom{n}{2}a^{n-2}b^2 + ... + \binom{n}{i}a^{n-i}b^i + ... + \binom{n}{n-1}a^1 b^{n-1} + \binom{n}{n}b^n,$$

where $\dbinom{n}{i} = \dfrac{n!}{(n-i)!\,i!} = \dbinom{n}{n-i}$, and in particular $\dbinom{n}{0} = \dbinom{n}{n} = 1$.

Remember, $0! = 1! = 1$ and $i! = i(i-1)(i-2)...(3)(2)(1)$.

Several former students of mine thought "3!" meant "!!**THREE**!!" but they were just being silly!

Example 1) Expand $(2+x)^5$.

Solution $(2+x)^5 = \dbinom{5}{0}2^5 + \dbinom{5}{1}2^4 x + \dbinom{5}{2}2^3 x^2 + \dbinom{5}{3}2^2 x^3 + \dbinom{5}{4}2 x^4 + \dbinom{5}{5}x^5$

$$= 32 + 5(16)x + 10(8)x^2 + 10(4)x^3 + 5(2)x^4 + x^5$$
$$= 32 + 80x + 80x^2 + 40x^3 + 10x^4 + x^5$$

Example 2) Find the coefficient of x^{15} in the expansion of $\left(2x^3 + \dfrac{1}{4x^2}\right)^{10}$.

Solution $\left(2x^3 + \dfrac{1}{4x^2}\right)^{10} = \displaystyle\sum_{i=0}^{10}\binom{10}{i}(2x^3)^{10-i}\left(\dfrac{1}{4x^2}\right)^i$

$\boxed{\text{Expand, using properties of exponents.}}$
$= \displaystyle\sum_{i=0}^{10}\binom{10}{i}\dfrac{2^{10-i}x^{30-3i}}{4^i x^{2i}}$ $\boxed{\text{Rewrite } 4^i \text{ in base 2.}}$ $= \displaystyle\sum_{i=0}^{10}\binom{10}{i}\dfrac{2^{10-i}x^{30-3i}}{2^{2i}x^{2i}}$

$\boxed{\text{Combine exponents!}}$
$= \displaystyle\sum_{i=0}^{10}\binom{10}{i}2^{10-3i}x^{30-5i}.$

We want $30 - 5i = 15$. $\therefore -5i = -15$ and so $i = 3$.

The required coefficient is

$$\binom{10}{3}2^{10-3(3)} = \dfrac{10!}{7!3!}2^1 = \dfrac{10(9)(8)}{(3)(2)(1)}(2) = 240.$$

Two for you.

1) Expand $\left(\dfrac{2}{x^2} - \dfrac{y}{3}\right)^4$.

2) Find the coefficient of x^{-4} in the expansion of $\left(\dfrac{3}{x^2} + \dfrac{x^2}{9}\right)^4$

Answers 1) $\dfrac{16}{x^8} - \dfrac{32y}{3x^6} + \dfrac{8y^2}{3x^4} - \dfrac{8y^3}{27x^2} + \dfrac{y^4}{81}$ 2) 12

HOW TO GET AN "A" IN MATH

1) After class, **DON'T** do your homework! Instead, read over your class notes. When you come to an example done in class...

2) **DON'T** read the example. Copy out the question, set your notes aside, and do the question yourself. Maybe you will get stuck. Even if you thought you understood the example completely when the teacher went over it in class, you may get stuck.

And this is **GOOD NEWS**! Now, you know what you don't know. So, consult your notes, look in the text, see your teacher/professor. Do whatever is necessary to figure out the steps in the example that troubled you.

Once you have sweated through the example, **DO IT AGAIN! And again**. Do it as often as you need so that it becomes, if not easy, then at least straightforward. Make sure you not only understand each line in the solution, but why each line is needed for the solution.

In part, you have memorized the solution. More importantly, you have made the subtleties of the problem unsubtle!

This is the great equalizer step. If your math or science aptitude is strong, then maybe you will have the example down pat after doing it twice. If not so strong, you may have to do it several times. But after you have done this for every class example...

3) **DO YOUR HOMEWORK!**

If you follow this method and if the teacher chose the examples well, then most of the homework questions will relate easily back to problems done in class and the rest should extend or synthesize the ideas behind those problems.

Guess what you'll find on 80% or more of your tests and exams? The same kinds of problems! And you will have your "A". Good luck, although if you use this method, luck will have nothing to do with your INEVITABLE success.

THE MATHEMATICS SURVIVAL KIT
FEEDBACK

What topic did you need that you didn't find in **The Mathematics Survival Kit**?

Send me your topic*. If I use it in a future edition of the Survival Kit, I will send you a copy of the new edition hot off the press. In it, you will find a page with your topic, and at the bottom of the page, **your name and school!** Be famous and help me improve the Kit. Send me your topic!

Good luck with your studies in general and mathematics especially.

Best wishes,

Jack Weiner
Associate Professor
Department of Mathematics and Statistics
University of Guelph
Guelph, Ontario, Canada

*Send your suggested topic to
The Mathematics Survival Kit
c/o Nelson Higher Education
Editorial Group
1120 Birchmount Road
Toronto, Ontario M1K5G4

INDEX